KIN

KIN

*How We Came to Know Our
Microbe Relatives*

JOHN L. INGRAHAM

Harvard University Press
CAMBRIDGE, MASSACHUSETTS
LONDON, ENGLAND
2017

Copyright © 2017 by the President and Fellows of Harvard College
All rights reserved
Printed in the United States of America

First printing

Many of the designations used by manufacturers and sellers to distinguish their products are claimed as trademarks. Where those designations appear in this book and Harvard University Press was aware of a trademark claim, then the designations have been printed in initial capital letters.

Design by Dean Bornstein

Library of Congress Cataloging-in-Publication Data

Names: Ingraham, John L., author.
Title: Kin : how we came to know our microbe relatives / John L. Ingraham.
Description: Cambridge, Massachusetts : Harvard University Press, 2017. | Includes bibliographical references and index.
Identifiers: LCCN 2016037619 | ISBN 9780674660403 (alk. paper)
Subjects: LCSH: Microorganisms—Evolution. | Bacteria—Evolution. | Evolution (Biology) | Life—Origin.
Classification: LCC QR13 .I547 2017 | DDC 579/.138—dc23
LC record available at https://lccn.loc.gov/2016037619

For Nancy

Contents

Preface ix
Introduction 1

Part One. Discovering the Tree of Life

1. The Tree's Microbial Branches 21
2. Relationships among Organisms 49
3. Enter DNA 80
4. The Rosetta Stone 97
5. From the Tree's Roots to Its Branches 116

Part Two. Doubts and Complications

6. Genes from Neighbors 141
7. Can the Receiving Cell Say No? 170
8. Can the Tree Be Trusted? 187

Part Three. Understanding the Tree of Life

9. The Tree's Ecological Fruit 217
10. The Tree's Beginnings 232

Further Reading 255
Acknowledgments 265
Illustration Credits 267
Index 269

Preface

In 1859 Charles Darwin published his paradigm-changing book *The Origin of Species by Means of Natural Selection, or the Preservation of Favoured Races in the Struggle for Life* and provided a rational explanation for the driving force of evolution. Biologists ever since have set about tracing evolution's multiple, interconnecting pathways with considerable success. By this means they have been able to outline relationships among organisms. Up until the 1970s all such studies were constrained by their reliance on a single kind of evidence: morphology, the shapes and sizes of things. Evolutionary relationships were traced by how organisms looked. Extant species and fossils of organisms that resembled one another were presumed to be related because they shared a common genetic past. Relationships among many plants and animals were quite successfully deciphered this way. Darwin himself participated. He traced the evolutionary history and relatedness of barnacles simply by meticulously studying the similarities and differences in the shapes of the parts of their various species. There are, of course, obvious limitations to such an approach. How do we determine the relationships among organisms that are so different that they lack any common morphological features? How is a pine tree related to a barnacle? In addition, how do we determine relatedness among a group of microorganisms whose morphology is so simple that many of them that we now know to be quite different genetically look almost the same. Or perhaps even more fundamentally, how do we determine the possible relationships of microbes to plants and animals? Distinguished biologists despaired that such questions could ever be answered. Then in the 1970s a scientific bombshell shook the field of biology. Biologists realized that

PREFACE

the history of relatedness of all living things is recorded in the molecular morphology of the constituents of cells. This insight has led to discoveries that have changed biology profoundly, allowing us to glimpse the entire universal Tree of Life and see for the first time our position in the complex web of nature, a revelation both humbling and revolutionary.

Come in; come in. The gods are here too.
—Aristotle, *On the Parts of Animals*

Introduction

We're fascinated by our origins. The world's vastly disparate religions share a common fascination with our and all life's beginnings. Both DNA testing and internet databases are available to search for one's relations, ancestors, and ethnic origins. We and our ancestors have often seemed driven to know where we came from as individuals and as a species. Relatively recently, some of these most primal questions have been definitively answered: we have come to discover our collective origins, at least in broad outline, and to recognize our remarkable kinship to microbes.

Modern biology has constructed a new scientific origin tale with an impact equivalent to cosmology's big bang theory. The components are the newly established relationships among living things. Assembled, they make a clear and dramatic statement: we are all kin. From the smallest bacterium to the largest blue whale, we are all fellow members of the same inclusive family of life. We are connected by common and traceable inheritance that leads back to this family's microbial beginnings.

Since Charles Darwin, biologists have suspected that we descended through a single, branching pathway of life, but now it's been unequivocally established. As we might anticipate, parts of the pathway have some unusual twists and turns. This common legacy of living things, which has been established with certainty only during the past forty-odd years, has gone largely unnoticed, or at least underappreciated, by those of us who aren't biologists. We still don't know where we're going and can only guess about how it all began,

INTRODUCTION

but now we do know with certainty where we and our fellow living creatures came from; to whom we are related and how closely.

This research establishes much more than life's singularity. It displays the relative closeness of the relationships among various organisms as well as their genetic interconnections. It tells how we acquired individual components of our cells: how our proteins evolved by assembling component parts from disparate sources, and how some of our essential cellular functions were acquired directly from microbes.

The new origin tale reminds us, as we'd already learned from traditional studies and from observing nature, that we're most closely related to other animals. As W. Ford Doolittle noted, before this achievement, "we would still find it easy to tell birds from bees, or distinguish any bird or bee from broccoli, brewer's yeast, or bacteria. But we would have no strong basis for deciding as we have that all birds and bees are closer kin to yeast than to broccoli." Now we know that our relationship extends to all of them. We share common ancestors with plants. Most surprisingly, perhaps, new discoveries not only firmly establish that we are descendants of microbes, they explain our connections to them. The most basic, life-sustaining activities of the cells comprising our bodies, including those that enable us to digest our food and derive metabolic energy from it, evolved in microbes. More fundamentally, we store our life instructions and pass them on to our progeny along the route pioneered by microbes. We weren't the venues for developing these fundamental bases of our being; we acquired them through inheritance from our distant microbial relatives.

Perhaps an even greater impact of these studies is the revelation that such things are knowable, that a question of such profundity may be answered by ordinary laboratory experiments, some seemingly mundane and tedious. Might other biological conundrums be similarly penetrable? Certainly the discovery of a totally new class of organism, the archaea, in 1977 came as a major surprise. These newcomers (to us) are unique among living things, as different genetically, biochemically, and evolutionarily from other microbes as they are from us. The novelty of their discovery is unparalleled in the history of modern biology—equivalent only to Anton van Leeuwenhoek's amazement upon discovering a world of microorganisms on the glass beneath his homemade, handheld microscope in the seventeenth century.

Throughout the history of biology, new organisms have been discovered at a steady rate—new species of beetle, spider, fish, plant, or bacterium—a flow that will undoubtedly continue or even accelerate as new detection methods become available, though species are also lost to extinction. But discovery of the archaea required the invention of a new term, *domain*. With its discovery the biological world became logically divided into three distinct domains: the archaea, the bacteria, and everything else, a category we call eukaryotes, which encompasses many other microbes, fungi, plants, and animals—including ourselves.

The Tree of Life

In letters to friends but never in publication, Darwin speculated about the possibility of a common ancestor. He wrote of plants and animals, but he never mentioned microbes, or "infusoria" as he called them, as possible candidates. Now we see that microbes are indeed ancestral to all plants and animals.

The Tree of Life, a term first used in a scientific sense by Darwin, summarizes the long-expressed speculations of biologists that there is a single evolutionary trail arising from life's beginnings that branches repeatedly and leads to all extant organisms. The tree analogy also expressed Darwin's hopes that a detailed map of life's evolutionary

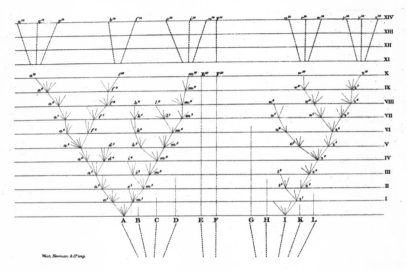

FIGURE 1. The only figure in Darwin's *Origin of Species* meant to illustrate how species at the top of the figure are related because they have a common descent from those shown at the bottom.

journey would one day emerge. Darwin's tree, the evidence for which seemed frustratingly unreachable at one time, reflects the essence of Darwin's concept of evolution—in his words, "descent with modification."

Of course, the Tree of Life was familiar in another sense to almost everyone in Darwin's time and long before: "And out of the ground made the LORD GOD to grow every tree that is pleasant to the sight and good for food; the *tree of life* also in the midst of the garden, and the tree of knowledge of good and evil." Indeed, this biblical tree from Genesis is central to Christian theology, but Darwin never acknowledged this rather different use of the term.

Darwin's epic masterpiece *On the Origin of Species* shook the entire world and continues to rattle some of it. (According to a 2014 Gallup

Poll, 42 percent of Americans believe that "God created human beings pretty much in their present form at one time within the last 10,000 years or so.") Long accepted in the scientific community, Darwin's book contains only one diagram in its first edition.

Darwin's simple line drawing of branching decent from a common ancestor leading to a set of cousin-like descendants illustrates how a particular group of similar varieties or species might have developed. The sketch had far-reaching and lasting impact. It was the speculative beginning of the Tree of Life, and it summarized the crux of Darwin's thoughts: *similar creatures appear to be similar because they are close relatives,* thus linking taxonomy (a system of naming) and phylogeny (evolutionary relationships). His diagram also illustrates how extant groups could be descendants of a now-extinct common ancestor. Darwin went even further in one of his notebooks.

He drew an actual treelike image: a single trunk branching at different points repeatedly, leading to extant organisms at the tips of the branches and extinct ones at the axillaries. Humbly, he wrote above it, "I think." Darwin made this sketch around July 1837, twenty-two years before he marshaled courage sufficient to publish his theory of evolution in the *Origin of Species by Means of Natural Selection, or the Preservation of Favoured Races in the Struggle for Life,* which questioned the religious convictions of many, including his pious wife. He knew it would offend. Darwin scholars agree he meant the tree to represent his conviction that all plants and animals share a common evolutionary history and a common ancestor. The sketch may even have reflected his hope that one day all species might be located on the tree. He wrote in *On the Origin of Species*: "The affinities of all the beings of the same class have sometimes been represented as a great tree. . . . As buds give rise by growth to fresh buds, and these, if vigorous, branch out and overtop on all sides many a feebler branch, so by generation I believe it has been with the great Tree of

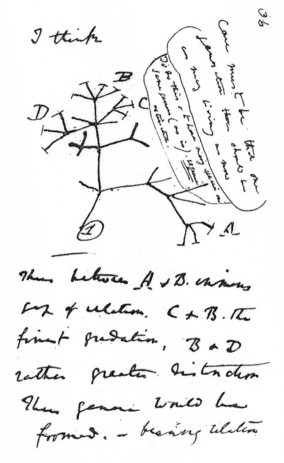

FIGURE 2. Darwin's concept of the Tree of Life.

Life, which fills with its dead and broken branches the crust of the earth, and covers the surface with its ever branching and beautiful ramifications."

Darwin scholars can't agree on whether Darwin thought microbes were our ancestors or even a part of this Tree of Life. Now we know

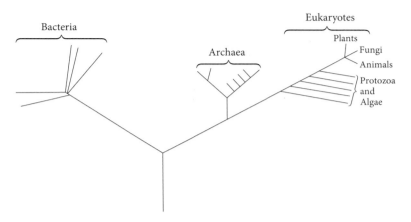

FIGURE 3. Arrangement of the major biological groups on the Tree of Life.

that they most assuredly are; they are the tree's beginning and its bulk. We know that they are responsible for most of the evolutionary innovations that we acquired from them through inheritance. We and our fellow macrobes constitute only a small set of twigs on one of the tree's uppermost branches. Darwin's use of the name *infusoria* for microbes, implying creatures that inevitably occur in an infusion made of meat or hay, suggests that he could have thought that microbes might not be part of the Tree of Life, that they appear spontaneously from nonliving material if conditions are favorable.

Darwin's speculative tree has a single major trunk from which branches sprout. We now know that the actual Tree of Life has three trunks, branching twice near its base.

Long before the Tree of Life was even imagined, humans named things, including living things, and grouped them. It's deep in our culture and nature. In the Bible, Adam names the creatures even before Eve appears: "and whatsoever Adam called every living creature, that was the name thereof." Taxonomy, the naming of things, and classification, the sorting of them into logical groups, have had a rocky relationship with evolution. We can name and classify most nonliving things without recourse to evolution, and we can and have classified living things with no consideration of evolution. An extreme example, perhaps, is that practiced by a group of sixteenth-century botanists who proposed classifying plants by the alphabetical order of their given names. But for most taxonomic schemes evolution lurked, even dominated. Similar-appearing organisms were grouped together, and morphological similarity usually derives from common evolution—but not always. A classic example of this exception, the Tasmanian wolf, is now extinct. Though it looked remarkably like a dog or wolf, it was a distantly related marsupial, not a placental mammal as all true canines are. Habitat and function as well as inheritance shape form. Evolution-based taxonomic schemes are more precise and offer the clear advantage of being derived from testable data.

Since Plato, most classification schemes, like his dialogues, were sequentially dichotomous: each branch of a pair of branches divided into two more. That seems to be our natural human inclination. Members of a group are either this or that, and then similarly we collect each of the two groups into two larger, more inclusive ones. Plato's dialogues never offered three choices. A series of sequential dilemmas led Socrates to the hemlock. (Since being introduced to him in high school I've often wondered why reaching such a patently ridiculous conclusion didn't cause Socrates to question his method of getting there.)

Likewise, throughout most of the history of biology, the living world was divided into two major groups, plants and animals. These divisions were titled kingdoms, as though they were ruled by an outside party. As new creatures, including microbes, were discovered and studied, they were assigned, sometimes quite awkwardly, to one or the other of the two major groups (taxons). Bacteria and fungi, because most of them for most of their life cycles don't move, were assigned to the plant kingdom; protozoa, which do move, ended up in the animal kingdom. The impact of this dichotomous style lingered even after it became scientifically and intuitively untenable. Until only a short time ago bacteriology and mycology, the studies of bacteria and fungi, were attached to academic departments of botany; protozoology, the study of protozoa, were assigned to departments of zoology. Assemblages of bacteria were called microflora (groups of small plants); now we call them microbiota (small living things) or, more recently, the microbiome.

The Tree of Life defies this human inclination toward bifurcation. Near its base, the tree splits twice, thereby generating three lines of evolutionary progression. These three major branches, termed domains, are the archaea, the bacteria, and the eukaryotes. Nature, apparently, did not follow or anticipate Plato's advice. We humans along with other animals, plants, protozoa, algae, and fungi all belong to the eukaryotic branch and sit atop its uppermost shoot. Together, we eukaryotic macrobes along with the fungi (although most fungi are microbes) constitute another tripartite set of branches—one so small as to be almost insignificant compared with the rest of the tree. Only plants, animals, and a few fungi are macrobes. The rest of the eukaryotic line and all of the two prokaryotic domains, bacteria and archaea, are microbes. We live as a minority in the vast sea of microbes, even within our own bodies. The number of microbial cells living on our body surfaces and in our gut is ten times greater than the number of our body's own cells.

The Path to the Tree

Relationships among living things have long fascinated humans. Over two thousand years ago Aristotle studied them and created a

INTRODUCTION

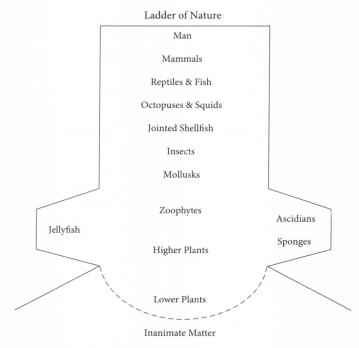

FIGURE 4. Aristotle's sketch of the Ladder of Nature, including life's beginning from inanimate matter.

scheme he called the Ladder of Nature. Although voluminous critical contributions to the quest have been made in the intervening years, it is only in the past forty years that the all-inclusive family of living things has been revealed.

That's the path we too will follow in this book: how Aristotle's ladder became the Tree of Life and how this tree was deciphered—the major events of its evolutionary past and concerns about its validity. In some ways, the path's beginning is startlingly prescient of its fruition. Aristotle clearly conceived of a biological progression from the simple to the complex, starting off with inanimate material, as later became shocking to some, and proceeding, as he said, "little by little from lifeless things to animal life . . . indeed, there is observed in plants a continuous scale of ascent toward the animal."

The noted historian of science Charles Singer speculated that had Aristotle lived another ten years, he would have developed the concept of evolution, or at least a sense of the relatedness of life forms, in spite of the paucity of information available to him. Aristotle certainly appears to have been close to such an intellectual breakthrough. But even if he had taken this monumental first step, he would have been a long way from constructing the Tree of Life. Observing the superficial similarity of organisms, particularly when it comes to their most abundant representatives, microbes, does not provide enough evidence to construct the tree. Its revelation, like all such major intellectual achievements, depended on an accumulation of a vast reservoir of information, a prepared mental state, and, of course, a few good ideas. It relied further on the availability of modern biochemical methods. As we follow the trail, we'll see many incremental steps of progress and only a few great leaps.

The path toward the Tree of Life is marked with a number of significant milestones, including such concepts as hierarchical schemes of relatedness and organic evolution, the rejection of the idea that new species continue to be generated spontaneously from inanimate matter, the recognition of continual genetic change through mutation, discovery of the genetic code and its mode of expression, as well as acknowledgment of the precision with which enzymes act on specific compounds.

New ways of thinking about life evolved often only after the old paradigm had been discarded. The idea of spontaneous generation, that under favorable conditions inanimate matter routinely becomes a living thing was long standing and long lived. Everyday observations seemed to affirm it. Rats frequently appeared in piles of discarded rags; maggots invariably developed in rotting flesh. The idea's longevity and persistence was remarkable, extending from Aristotle in about 343 BCE well into the late nineteenth century. In his *On the Generation of Animals,* Aristotle collected the observations of other scholars and proclaimed unequivocally, "Therefore living things form quickly whenever this air and vital heat are enclosed in anything." His conviction was later supported by the Bible, which in the book of Genesis proclaims, "Let the waters bring forth abundantly the moving creatures that hath life." Belief in spontaneous generation survived the rationality of the Renaissance and the Age of Reason.

The demise of the idea of spontaneous generation proceeded methodically from the larger to the smaller, from macrobes to microbes, through thoughtful experimentation. In 1668 Francesco Redi led the movement. Redi, a physician, naturalist, and poet, was a contemporary of Galileo. He showed that the seemingly inevitable generation of maggots in decaying meat could be stopped by the simple expedient of covering it with fine Naples veil. The maggots then appeared on the covering, where flies had laid their eggs, rather than in the meat.

But what about microbes? Might they be an exception to the rule that organisms come only from parents? After all, microbes are simple creatures. Nineteenth-century thinkers considered bacteria, for example, to be mere bags of protoplasm, small bits of life. Again, common experience seemed to offer powerful support for spontaneous generation. Stored perishable items soon teem with microbes. Within days sparklingly clear infusions of meat or hay (extracts) inevitably become cloudy owing to the growth of huge numbers of microbial cells.

A series of respected scientists—including Pier Antonio Micheli, John Needham, and Lazzaro Spallanzani, attacked the question of microbes' sudden appearance with mixed and conflicting results.

INTRODUCTION

In the early eighteenth century Micheli, a Roman Catholic priest from Florence, made perhaps the most sophisticated, scientifically founded arguments against the spontaneous generation of fungi. He showed that fungi develop from microscopic spores, and not until such spores are placed on a cut melon do fungi develop, causing the melon to rot.

Somewhat later in the eighteenth century John Needham, the first Roman Catholic priest to be elected to the Royal Society, confused the issue with his misunderstanding of the limits of sterilization. Needham made various extracts of natural materials and boiled them, a treatment he thought adequate to kill all life. He then sealed them in containers, but the contents spoiled nevertheless. Needham used his results to illustrate the concept of vitalism, the idea that there was something quite special about life that made it resistant to physical laws. Life could, he believed, generate spontaneously when conditions were right.

Soon thereafter Lazzaro Spallanzani, an Italian priest who made important contributions to our understanding of animal reproduction, shifted opinion again. He repeated Needham's experiments, but his infusions did not spoil.

These arguments continued well into the nineteenth century. In 1855 a prestigious advocate of the idea of spontaneous generation of life made a proposal. Félix Pouchet was director of the Natural History Museum in Rouen. In a series of papers he offered an idea that seemed highly plausible at the time: it's part of God's plan. He theorized that eggs, the origin of all animals, are single cells that are spontaneously generated. So are microbes, by the same means. As proof, Pouchet prepared clear extracts of hay and heated them, sufficiently he believed, to kill any microbes that might be present, then sealed the containers. Routinely, the extracts became cloudy within a matter of days, owing to the presence of huge numbers of microbial cells.

INTRODUCTION

Then Louis Pasteur, already an established and highly respected luminary for his many achievements, both scientific and practical, entered the fray. Pasteur was motivated by the 2,500 franc Alhumbert Prize offered by the French Academy of Science to whomever could shed new light on the question of spontaneous generation. In 1862 the prize was awarded to Pasteur for his "Mémoire sur les corpuscules organisés qui existent dans l'atmosphère" published in the Academy's *Annales de chimie et de physique*. Among his experiments were those with the famous swan-necked flasks. He put clear meat broth into these flasks, bent the necks of the flasks into an S-curve, and sterilized the contents by boiling. Thus the contents were open to the air, which the proponents of spontaneous generation considered essential to the process, but sheltered from dust particles, which Pasteur believed to be carriers of microbial cells. These particles would settle in the trough of the flask's swan neck. The scheme worked. The broth remained clear. Indeed, one of his flasks, so prepared, resides in the basement of the Pasteur Institute in Paris, still microbe-free after some 150 intervening years. To support his rationale, Pasteur poured a portion of the contents of some flasks into the trough in the neck and then poured it back into the flask, thereby washing settled dust particles with their presumed microbial passengers back into the flask. These flasks soon clouded with microbial growth.

In June 1864, in order to resolve their differing results, a public competition was arranged for Pasteur and Pouchet to demonstrate their methods. Claiming bias, Pouchet withdrew, fortunately for Pasteur because the results of their experiments almost certainly would have conflicted. Pasteur prepared his broth from meat extracts; the bacteria in them are readily destroyed by boiling. Pouchet's infusions, however, were made from hay, which almost always contains highly heat-resistant bacterial endospores, which can survive boiling. The outcome of the Pasteur / Pouchet controversy is another of the

INTRODUCTION

FIGURE 5. One of Pasteur's swan-neck flasks.

reasons that Pasteur is called a "precarious genius," and for the most part it put to rest belief in the frequent, contemporary spontaneous generation of new life forms.

In the mid-1870s John Tyndall, an Irish-born physicist working in Scotland, incisively tied up the loose ends of the controversy. Tyndall was already a distinguished scientist who had made important contributions to our understanding of diamagnetism and infrared radiation. He was also interested in the thermal properties of air. Tyndall observed that air always contained minute suspended particles, which he could observe in a darkened chamber because they scattered a beam of light that was visible when observed at a right angle to the light source, a now familiar phenomenon known as the Tyndall effect. He became curious about the properties of particle-free, "optically pure" air, which he obtained by coating the inside of a box with glycerin, reasoning that the particles would be trapped in a layer of the sticky liquid as they randomly collided with its walls. It seemed to work. A few days later, a strong beam of light passing through the box revealed no light-scattering particles. He then placed open test tubes of urine and various meat extracts in the chamber. These remained clear, whereas those placed in a similar chamber with ordinary air routinely became cloudy after a few days, a direct confirmation of Pasteur's suspicion that ordinary air is filled with microbes, either residing on dust particles or floating freely.

Tyndall then addressed the issue of the hay infusions. He boiled the hay infusions on three successive days, reasoning that in the interim, heat-resistant spores in the infusion would germinate, producing vegetative cells susceptible to heat. It worked. Hay infusions so treated remained clear when exposed to "optically pure" air. This three-day treatment for sterilizing materials that contain heat-resistant bacterial endospores became known as tyndallization. Tyndall's and Pasteur's experiments put an end to belief in present-day spontaneous generation. With the demise of this concept, scientists were much closer to an understanding of descent, essential to the concept of the Tree of Life. What Pasteur, Tyndall, and their predecessors had established is that life does not arise readily and rapidly under extant conditions. However, life at its beginnings must have arisen from nonliving matter at a time in which conditions on Earth were quite different. In the final part of this book we will consider the roots of the tree—how life might have begun.

∽

The big idea leading to the revelation of the Tree of Life was the recognition of the vast reserves of historical information held within all cells, encrypted there in the sequence of the components of certain of a cell's macromolecules: proteins, RNA, and DNA. Biologists came to

INTRODUCTION

realize that sequences of the amino acids, the strings of them linked together to make up proteins, and the strings of nucleobases making up a molecule of RNA or DNA all encode a detailed and complete history of evolution. It is as if modern-day archaeologists had come upon a vast, inexhaustible mine of fossils dating back to Earth's origin. The goal of microbiologists has been to develop the means of excavating this mine of information—precise methods for separating and identifying quite similar molecules as well as the sophisticated computer technology needed to process huge volumes of data.

{ PART ONE }

DISCOVERING THE TREE OF LIFE

CHAPTER 1

The Tree's Microbial Branches

Even a quick glance at the Tree of Life emphasizes a startling fact: it's composed mainly, almost exclusively, of microbes. Plants and animals, although they dominate our visual world, constitute only two of the tree's branches. The bacteria, archaea, protista, and most of the fungi are microbes. In spite of their microbial commonality, these microbes differ dramatically in their structure and activities.

The Fungi

The Tree of Life identifies fungi as our closest microbial relatives; slightly more distant branches are the protozoa and algae (collectively termed protists). Next follow the two massive domains of prokaryotes, those very different microbes, the archaea and the bacteria, that share a strikingly simple cell architecture lacking internal membranes, even a defined nucleus. That's the basis of the name: *pro-*, in place of or before; and *karyote,* or nucleus. Cells of prokaryotes appear to be densely packed with a uniform, undifferentiated, granular substance. Even a quick microscopic glance is enough to distinguish a prokaryotic cell from all others. . Fungi—the mushrooms, molds, and yeasts—are surprisingly close relatives. They're closer cousins to us and other animals than they and we are to plants.

Cell biologists have capitalized on our close connection to fungi by focusing on yeast to probe fundamental biological questions. Yeasts are easy to cultivate. They grow relatively rapidly and have genetic properties that facilitate gene manipulation. Thus, they are useful for probing fundamental biological questions relating to longevity and the intricacies of gene expression. Yeast is a loose, general term, referring to all fungi that grow for most of their life cycle as single cells. One particular yeast has become the central focus of these studies—a model organism second only perhaps to the intensively investigated bacterium, *Escherichia coli*. When we say "yeast," we almost always mean *Saccharomyces cerevisiae*.

Humans have shared a long and intimate history with *S. cerevisiae*. When it ferments simple sugars such as glucose or fructose, it produces carbon dioxide and ethyl alcohol, both of which have been exploited by humans. In the presence of air, yeast respires (metabolically "burns") sugars as we do; in its absence it ferments. Louis Pasteur had yeast in mind when he famously defined fermentation as "life without air." (We now know that fermentation is not the only way that microbes can gain energy for life in the absence of air, but it is the way that yeast and many other microbes do.) A more accurate, if less colorful, definition of fermentation is splitting a nutrient and using one fragment to oxidize (take electrons away from) the other, thereby gaining metabolic energy, or more generally by passing electrons between two organic molecules. Since antiquity, humans have exploited the carbon dioxide released as yeast ferments sugar to leaven bread and make sparkling beverages. Without the alcohol produced by fermenting yeast, wine, beer, liquors, and even some fuels would not exist.

Making wine is relatively simple. It is the inevitable consequence of crushing grapes. All of the necessary ingredients are present. Yeasts are almost always present on the skin of the grapes in the powdery layer called "bloom," and grapes' internal sweetness depends on an equal mixture of glucose and fructose, two simple sugars. When the grape is crushed, the yeast comes in contract with these sugars and ferments them, releasing carbon dioxide and ethanol. Winemaking is essentially automatic, though achieving consistent, palatable excellence requires a more controlled approach.

Making beer and liquor is not so simple. These are made from grains, which contain starch, not sugar, as stored nutrients. Yeast cannot ferment starch. So making beer or the mash from which whisky and other hard liquors are distilled requires human intervention prior to fermentation. The grain's large starch molecule must be disassembled into its component sugars, glucose and maltose (a combination of two glucose units), a process called saccharification. Most human societies discovered this process in antiquity. There is abundant, solid evidence that ancient Egyptians and Chinese knew about saccharification. The essence of saccharification is finding and adding a source of amylase, the enzyme that breaks down starch into its component simple sugars.

The Egyptians may have been the first to use malt as a source of amylase for saccharification. Malt is barley that has been dried just as it begins to germinate. As grains begin to germinate, they produce amylase to convert their own stored starch into glucose to nourish their embryos as they start to grow. Asian cultures use fungi to produce the amylase needed for traditional beverages such as sake and Chinese rice wine. A mixture of filamentous fungi, largely *Aspergillus oryzae*, provide a source of amylase. The fungi also confer a distinctive flavor. Human saliva is also a rich source of amylase, one that Native American and African cultures have used to make their traditional beverages. Desire to make alcoholic beverages drove human innovation worldwide, for better or worse.

Of course, yeasts constitute only a small fraction of the world of fungi. In contrast to yeast, most fungi are not single cells; they are filamentous, forming long, tubular structures. Fungi are everywhere and do many things. They are ubiquitous in the soil, where they are the major degraders of organic matter. They are probably the main cause of plant diseases, and they infect some animals as well. Because of our close relationship to fungi, the systemic diseases they produce, such as coccidioidomycosis (San Joaquin Valley fever) are notoriously difficult to treat with antimicrobial drugs. The targets for their antifungal action are generally so similar to our own that antifungal drugs often cause significant side effects.

On the other hand, the fungi also include mushrooms, which enrich our lives. Think of life without porcini.

The Protists

In addition to the fungi and the two prokaryotic domains, there is a huge group of microbes that is a bit more distantly related to us than the fungi. These were traditionally called protozoa and algae, but are now more frequently grouped under the umbrella term *protists.* These microbes share with plants, animals, and fungi a cell architecture containing a "true" nucleus, a double-membrane-bound structure in which the cell's DNA is stored (thus *eu-,* true; *karyote,* nucleus). This type of cell architecture defines protists (and fungi, plants, and animals) as eukaryotes. Protists are an enormous, ubiquitous, and diverse group of microbes that includes such notorious human pathogens as those causing malaria, sleeping sickness, and amoebic dysentery. Various protists are abundant in almost any aqueous environment, from mud puddles to oceans. You may have seen them scurrying around in the pond water you examined through a microscope in biology class.

Photosynthetic protists play a massive ecological role, particularly in the oceans, being responsible for a major portion of atmospheric carbon dioxide that is fixed, or made available to organisms. One group of photosynthetic protists, the diatoms, has left geological evidence of their abundance. Deposits of diatomaceous earth or diatomite are composed of layers of diatoms that have left their glasslike cell walls made of silica (SiO_2) behind. These layers accumulated on the bottoms of ancient seas, principally during the Miocene and Pliocene Ages, 2 to 25 million years ago. The largest is probably the Sisquoc Formation near Lompoc, California, which is 5,000 feet thick. The stunningly beautiful, geometrically precise cells that make up these layers are important articles of commerce. They may be used as an additive to concrete, food, cat litter, metal polishes, and toothpaste. Diatoms help with filtration, form a stabilizing component of dynamite, and kill most beetle-type insects, including bed bugs and fleas. Their minute shards pierce the insect's shell and puncture its body.

Another group of photosynthetic protists, the coccolithophores ("round stone bearers"), has also left a substantial geological footprint on Earth. These single-celled organisms are characterized by their signature sculptured exoskeletons, which are made of calcium carbonate ($CaCO_3$). These skeletons are the major component of chalk. Coccolithophores are still abundant in our oceans today, but they flourished particularly during the Cretaceous Period (145 to 66 million years ago), when concentrations of atmospheric carbon dioxide were elevated and as a result the oceans were warm and acidic, covering much of the globe. Parts of what are now Europe and the United States contained inland seas. The exoskeletons of these coccolithophores accumulated on the ocean bottoms, forming a layer of chalk that in places reached a thickness of 1,200 to 3,000 feet. The chalk layer, the defining characteristic of the Cretaceous Period (from *creta,* or "chalk") occurs worldwide. Portions of it are exposed in a number of places, most famously at the White Cliffs of Dover—perhaps the origin of "Albion," the oldest-known name for Great Britain. Views of an exposed layer of chalk may be seen in France on Normandy's "Alabaster Coast" and in Møns Klint, a six-kilometer stretch of cliff face on the island of Møn in Denmark. The chalk layer of the Cretaceous Period is a huge repository of carbon dioxide, trapped there as atmospheric gas dissolved in water to form carbonic acid (H_2CO_3), which coccolithophores combined with calcium ions to form their stonelike exoskeletons of calcium carbonate.

The chalk layer made by coccolithophores was the theme of Thomas Henry Huxley's powerful support of evolutionary theory, "On a Piece of Chalk," published in 1868 in *Macmillan's Magazine*. Some have called this short essay the finest piece of scientific writing ever. Huxley, who earned the epithet "Darwin's Bulldog," begins by recounting the extent and depth of the subterranean chalk layer in Britain from the observations of well diggers. He then describes microscopic observations showing that the major component of the layer is the skeletal remains of coccolithophores, which he calls *Globigerina*. These were living creatures of the sea, he concludes, noting that similar microbes are present in the Atlantic Ocean and have accumulated there in smaller amounts on the sea floor. An array of fossils of other organisms lie embedded in the layer of chalk, for example, a coral growing on a clam shell, and with this evidence Huxley is able to estimate the time required for the chalk layer to form and when it might have occurred on the evolutionary scale. He finds evidence of land plants above the chalk layer and then more sea dwellers. He concludes that Earth is in a constant state of flux: the sea is thrust up to become land and is again inundated, although he has yet no idea of plate tectonics, the driving force that has been only recently discovered. He then discusses the differences and relatedness of animal species above and below the chalk layer, concluding it improbable that they are the products of separate acts of creation. One must have evolved into the other. All this is derived from a piece of chalk and Huxley's penetrating reasoning. Of all the references cited, I recommend "On a Piece of Chalk" most highly. It's a delight.

The Archaea

Of the two prokaryotic microbial domains, microbiologists were surprised to learn, we have a more intimate evolutionary connection to the archaea, that relatively recently discovered domain of life represented by such unusual microbes as those that can make natural gas (methane) from two other gases (hydrogen and carbon dioxide) and those that can live in a variety of extremely hostile environments, including blistering hot springs with temperatures as high as 114°C (237°F—well above the boiling point of water), saturated brine, or locales as corrosive as battery acid. With their astounding skills for colonizing harsh environments, many archaea dominate the realm of microbes called extremophiles, organisms that flourish in environments that most living things would find inhospitable. The discovery of archaea has pushed our concept of the limits of life-sustaining habitats to the near unimaginable. Some archaea tolerate or even need what most organisms would find lethal. A prime example is the recently discovered *Pyrococcus yayanosii*, a resident of the blistering "black smokers" emanating from the ocean floor, two and a half miles deep, at the Mid-Atlantic Ridge, collected there by a submersible drone operated remotely and electronically from the wonderfully named research vessel, *Pourquoi Pas?* This French marine research ship was named after the ship of the polar explorer Jean-Baptiste Charcot, which was lost off Iceland in 1936.

P. yayanosii has bizarre preferences for punishing insults of high hydrostatic pressure, temperature, and concentrations of salt. It can tolerate and thrive at the crushing pressure of 11,843 atmospheres (174,044 pounds per square inch), although it prefers a more moderate hydrostatic pressure of a mere 513 atmospheres. Indeed, it needs an intensely high-pressure environment to live. It can reproduce only when subjected to at least 247 atmospheres. It cannot multiply at 1 atmosphere, a pressure comfortable for humans and most of Earth's inhabitants. It has similarly startling tastes for elevated temperature, tolerating and multiplying at 108°C (226°F), not a record holder but close to it, although preferring a slightly more moderate 98°C (just below the boiling point of water at sea level). It cannot multiply at all at temperatures less than 80°C (176°F), ten degrees higher than a well-roasted turkey and a lethal temperature for most organisms—including most microbes. Thus, the archaeon *P. yayanosii* is obligately piezophilic (pressure loving) and hyperthermophilic (extreme high-temperature loving); it thrives in an environment that would instantly crush and cook most living things. Moreover, this archeon tolerates and flourishes in a very salty environment of 3.5 percent sodium chloride, more salty than present day seawater (3.0 percent).

P. yayanosii has its phobias as well as its bizarre preferences. It cannot tolerate oxygen; even a brief exposure to air is lethal. This peculiar microbe derives energy from available organic nutrients by "burning" chemically oxidized forms of sulfur instead of oxygen. *P. yayanosii* is by no means a typical archaeon, but it dramatically illustrates the penchant of some of them for environmental extremes. There are many such organisms among the archaea, some of which can be seen in such unlikely environments as the hot springs in Yellowstone National Park.

Some groups of archaea prosper in the presence of high concentrations of salt. Many of these are red; they give salt-rendering ponds their characteristic red color. As these ponds become increasingly salty, successive populations of increasingly salt-tolerant archaea appear in them. A curious and distinctive archaeon, *Haloquadratum walsbyi*, often develops in the saltiest ponds; its cells are bright red, thin, and square.

The Tree's Microbial Branches

Although the archaea dominate extreme environments, they are not limited to them. Archaea are abundant in such moderate locations as soils, swamps, sewage, and oceans. About 20 percent of the ocean microbiota is archaea. They are even present in the intestinal tracts of humans and other animals.

The existence of archaea as a distinct biological group, let alone our relationship to them, is a quite recent discovery. Until the mid-1970s, some of the microbes that later were recognized as being archaea were thought to be just another group of bacteria; after the realization of their uniqueness, many other representatives have been discovered. Distinguishing archaea by examining their morphology is useless. They look like bacteria: their cells, on average about thirty-five times smaller than ours, are about the same size as those of most bacteria, and they share with bacteria that simple cell architecture called prokaryotic, lacking internal membranes, even a defined nucleus. But in spite of sharing prokaryotic cellular architecture with bacteria, the archaea in other respects are spectacularly different from bacteria and from all other creatures as well. Their biochemical distinctions are astonishing. For example, their version of the plasma membrane, the enclosing sack that surrounds all cells, is composed of chemicals found nowhere else in nature; the chemical composition of some of their cell walls is likewise unique; certain of their enzymes bear no structural resemblance to those serving the same function in bacteria or eukaryotes; some, as we've seen, thrive in environments that no other creatures can tolerate.

Moreover, there's a package of seemingly trivial distinctions between archaea and bacteria that suggest fundamental evolutionary divergences. For example, both motile archaea and bacteria swim by rotating stiff corkscrew-shaped proteinaceous flagella, but these develop quite differently. Those of bacteria elongate by adding subunits to their tips; those of archaea insert subunits at the flagellum's base, where it attaches to the cell.

In startling contrast to many bacteria and eukaryotic microbes, no archeon is known to cause disease in humans or any other creatures. This is fortunate because archaea are uniformly resistant to almost all antibiotics. Some microbiologists have reasoned that archaea appeared before other organisms existed, so there was no selective advantage or even opportunity for them to cause disease. In any event, the archaea constitute a totally distinct group of living things. They are as distantly related to bacteria, their fellow prokaryotes, as they are to us. The discovery of archaea as constituting a discrete group of organisms led directly to revealing the complete Tree of Life, the branching family tree of living things.

The Bacteria

You may know bacteria, the other and more familiar domain of prokaryotes, as "germs," the target of household cleansers and source of disease. But these microbes, our somewhat more distant kin, also provide many benefits to our planet. The Tree of Life reveals our direct connection to bacteria, as well as their relationships to one another and to the archaea. Although we've known of their existence since 1677, when Antonie van Leeuwenhoek, the Dutch draper and microscopist, first saw and described them to the British Royal Society, our relationship to them and their relationships to one another remained an impenetrable mystery until recently.

Bacteria constitute a huge group of diverse organisms. Although the number of recognized species is fewer than that of certain eukaryotic assemblages such as insects, they are noted for their diversity. Their impact on us and the environment is profound. Two evolutionary innovations of bacteria are particularly noteworthy: one group of them, the cyanobacteria, developed the capacity for oxygen-yielding photosynthesis, the process upon which all higher forms of life depend for their existence. And a major portion of bacteria, the Gram-negative bacteria, evolved to be surrounded by a double set of membranes, which greatly increased their potential to survive chemical assaults, a property that lives on in the mitochondria that almost all eukaryotic cells contain.

Bacteria as well as archaea have traditionally been studied by isolating them—separating a strain of the microbe from its natural environment and cultivating it in a laboratory by supplying quantities and kinds of nutrients that the investigators think should be adequate. This technique of studying pure cultures has been extraordinarily successful. It is the way that almost all the bacteria that cause infectious diseases in humans were identified in the late nineteenth century and have been studied since. This approach works very well for many bacteria, fortunately including most of those that cause infectious disease. But it has become increasingly clear over the past several decades that such an approach just doesn't work for the vast majority of bacteria. The number of bacterial cells that can be seen in a sample of soil, for example, is far fewer than the number of strains that can be isolated from that sample and cultivated in the laboratory. Using modern methods, the amount and diversity of bacterial DNA in an environment can be studied directly without cultivation in the laboratory. Such studies have encouraged microbiologists to expand their thinking about bacterial diversity. It is now more apparent than before that bacteria overwhelming dominate the Tree of Life.

The Tree's Phantom Branch

Viruses are not included on the Tree of Life. They lack the property that most fundamentally defines living things: they are not cells. They lack a means of generating their own metabolic energy, a means of self-reproduction, and ribosomes, among other cellular characteristics. In spite of not qualifying for representation on the Tree of Life, viruses are biological entities that can hardly be ignored in any discussion of the tree, if for no other reason than their staggering abundance and impact on the tree's legitimate occupants. Curtis A. Suttle makes this point dramatically with respect to the viruses in oceans. "Viruses are by far the most abundant 'lifeforms' in the oceans and the reservoir of the genetic diversity of the seas. The estimated 10^{30} viruses in the ocean, if stretched end to end, would span farther than the nearest 60 galaxies. Every second, approximately 10^{23} viral infections occur in the ocean. These infections are a major source of mortality, and cause disease in a range of organisms from shrimp to whales."

The Nobel Prize–winning biologist Peter Medawar succinctly defined what he considered a virus to be: "a piece of bad news wrapped up in a protein." That's its basic structure: just nucleic acid–encoded genetic information—some form of RNA or DNA—enclosed in a protein capsid. Some viruses do pack an enzyme or two within the capsid, and some are surrounded by a membrane that they stole from the last cell they infected. As these membrane-enclosed viruses bud from the cell, they become wrapped in a bit of that cell's membrane.

All viruses share another deficiency that separates them from the vast majority of living things. They can't reproduce by themselves. Instead, they depend on some living thing to do that for them. All living things at their own peril unwillingly participate in this massive biological conspiracy. There's no known exception to the rule that all organisms are susceptible to being infected by some virus, usually by several or many, thereby producing more viruses.

Viruses were introduced to the known world of biology only quite recently, in the late nineteenth century. Of course, some viral diseases had been studied long before that, but causes of such diseases were unknown. In 1798 Edward Jenner successfully developed a vaccine against the viral scourge smallpox. In 1885 Louis Pasteur successfully protected Joseph Meister against the virus causing rabies using a vaccine he had developed. Pasteur, who had developed the germ theory of disease, was unable to find the germ that caused rabies. He speculated, presciently of course, that the agent was too small for him to see through a microscope.

Size proved to be the first defining property of viruses. Charles Chamberland, one of Pasteur's colleagues, should probably be given a large portion of the credit for their discovery. He realized that pores in unglazed porcelain were smaller than bacteria, and a bar of it could be used to filter bacteria out of a liquid, thereby sterilizing it. Such filters became useful tools for microbiologists. (They were still in use early in my career.)

Almost simultaneously in the 1890s, two microbiologists, Dimitri Ivanovski in Russia and Martinus Beijerinck in Delft, Holland, discovered that extracts that were capable of transmitting mosaic disease from one tobacco plant to another could not be sterilized by passing them through a Chamberland filter. Ivanovski suspected that the infectious agent was a bacterium small enough to pass through the filter. Beijerinck suspected that it was something totally different, different enough to deserve its own new name. He chose *virus,* from the Latin meaning "poison." The agent of tobacco mosaic disease and others that were soon discovered to pass a Chamberland filter became known as "filterable viruses." Over the years the modifier was dropped.

Not until the twentieth century, with the advent of the electron microscope, could the array of viral shapes and sizes be seen. Some, like tobacco mosaic virus, are long and rod shaped. Others, like HIV, are spherical, and a few, like T4 phage that infects *Escherichia coli,* are complex, with an icosahedral head attached to an elaborate tail sporting fibers like insect legs and terminating in a spiked plate. Sizes of viruses also vary enormously. At one extreme are some that infect plants. They consist simply of a few hundred nucleobase, circular, single-stranded RNA, which because they lack a protein capsid are called viroids. At the other extreme are mimivirus and mamavirus, almost a micron in diameter, making them as large as some bacteria and thereby defying their earliest definition.

Perhaps the greatest variation among viruses is in the structure of their genes. It seems that any form of nucleic acid is possible in a virus. It can be composed of RNA or DNA. It can be segmented or a single molecule. The molecules can be circular or linear. Both the RNA and DNA can be single stranded or double stranded. If it is single stranded, the RNA can be the sense strand (the one that is directly copied or translated) or the antisense strand (the one that must be converted into a sense strand to direct synthesis of a protein).

Some virus particles contain one or more enzymes in addition to genes inside their capsid. In general they contain only those enzymes that their host cannot supply for them to begin to proliferate. Some of these special enzymes have been exploited successfully as targets for antiviral drugs, such as the enzymes associated with human immunodeficiency virus (HIV), which causes AIDS. HIV contains three enzymes within its capsid. One, called reverse transcriptase, converts RNA into DNA. This is a required transformation because HIV's genetic information is stored as RNA in the capsid but replicates as DNA within the infected cell. The HIV capsid also contains a protease, which is essential because the HIV genome directs its host to make a single long protein, which its protease cuts to form the virus's three essential enzymes. The drugs that have been so successful in treating AIDS inhibit the activity of HIV's reverse transcriptase or its protease.

Despite being left out of the Tree of Life, viruses have been named and classified by virologists into families, depending on similarities of their morphology and composition of their genomes. But, of course, viruses don't have ribosomes, which are so important to the tree's structure. At present, there's no way to way to relate viruses to the tree.

How viruses came to be and how they relate to life remains a mystery. There are two current theories. One holds that viruses evolved early in the history of life, possibly before cells had appeared. The other proposes the opposite, namely, that viruses are products of evolved cells, or "genes on the loose," the ultimate selfish gene. The latter theory, the one I prefer, holds that certain genes, escapees from an organism's genome, became encased in a capsid and by selection became able to multiply at the expense of another cells.

The Impact of Microbes

Microbes and their many evolutionary adventures constitute the vast majority of life's experiments with diversity and innovation. They made the major evolutionary leaps in genetics, metabolism, and cell structure that we acquired from them through inheritance. We tend to neglect our kinship with microbes because we can't see them, but they molded our evolution, our environment, and much of our planet's development as well. Their activities continue to moderate the damage we've done to it.

Earth's atmosphere is a good example of their profound impact. The components of the major portion of Earth's atmosphere, for example, are the result of the activities of microbes and are largely maintained by them. Microbes are responsible for making all the oxygen and nitrogen gases that air contains. Thus, microbes are responsible for making over 99 percent of the atmosphere we live in. Plants, of course, produce vast quantities of oxygen, but biology's studies on relatedness have firmly established that they can do this only because their ancestors captured and they have long since maintained bacteria as intracellular slaves, which in their evolved form are called chloroplasts. Chloroplasts photosynthesize for plants, spewing oxygen into the atmosphere as a by-product of this light-driven, energy-capturing process so fundamental to Earth's ecology.

In contrast to oxygen, atmospheric nitrogen is supplied exclusively by free-living microbes. Microbes, without assistance from any other creatures or processes, produced all the nitrogen gas in our atmosphere, and they continue to replenish it as portions of Earth's atmospheric reservoir are continuously converted to ammonia (NH_3) and subsequently other nongaseous forms of nitrogen used by plants. This conversion process is more commonly called "fixing" nitrogen. The cycling of nitrogen through its various nongaseous chemical forms is the exclusive realm of microbes: some convert the fixed ammonia into nitrite (NO_2^-); others further convert it into nitrate (NO_3^-); still others transform the nitrogen in nitrate back to its gaseous form (N_2). Until about a hundred years ago, microbes alone were responsible for almost all of the nitrogen gas taken from Earth's atmosphere. About 10 percent of the total fixed nitrogen was mediated by lightning strikes. Microbes thus supply fixed nitrogen (ammonia and nitrate) to plants and algae, which make many cellular components, including proteins and DNA, from this nutrient.

Nitrogen fixation, until humans intruded in the process in the early twentieth century, was the province of prokaryotes, both bacteria and archaea. This capacity is widespread among them. About 15 percent of prokaryotes that have been sequenced carry genes encoding the mediating enzyme, nitrogenase, and microbes have been fixing nitrogen for a very long time. Isotopic preference examinations of ancient rocks have established that biological nitrogen fixation began about 3.2 billion years ago, well before the Great Oxidation Event, when microbes flooded our atmosphere with oxygen.

Nitrogen fixation has two highly distinctive and metabolically unusual characteristics. It demands extraordinary amounts of energy—

about fifty molecules of ATP are needed to fix a single molecule of nitrogen gas—and it's sensitive to oxygen. Nitrogenase, the crucial enzyme of the process, is inactivated by even a brief exposure to the gas. Some suggest that the sensitivity of nitrogenase to oxygen reflects its ancient origins, that it evolved in an oxygen-free world. There's a logic to that argument. A source of fixed nitrogen must have been essential to life since its emergence. Owing to the unusual properties of nitrogenase, nitrogen fixation can occur only if an abundant source of energy is available and all oxygen is excluded. In spite of these severe limitations, prokaryotes have evolved various schemes to fix nitrogen in a variety of environments. The simplest is mediated by certain members of the bacterial genus *Clostridium*. Its members acquire abundant energy by fermenting carbohydrates in the complete absence of oxygen. Another bacterium, *Azotobacter,* creates its own oxygen-free environment for nitrogen fixation. It grows in the presence of air but uses aerobic respiration to generate energy so vigorously at the cell's periphery that its core is sufficiently free of oxygen for nitrogen to be fixed. These free-living bacteria, however, are the exceptions.

Most biological nitrogen fixation occurs as a collaborative venture between bacteria and plants, the plant providing the source of energy and the low-oxygen environment, the bacterium doing the actual fixing of nitrogen. The elaborately choreographed symbiosis between members of the genus *Rhizobium* and related bacteria with legumes—peas, beans, and alfalfa with their characteristic blue-bonnet flowers—is an interesting example. These bacteria can live and multiply freely in soil, but they don't fix nitrogen there. However, if a leguminous plant is nearby, one constantly releasing chemical attractants, these bacteria are enticed to swim towards the plant's roots. When an individual bacterium arrives at the root, it attaches to one of the root's single-cell extensions, which are called root hairs. The plant then

FIGURE 6. A pea plant with nitrogen-fixing nodules on its roots.

welcomes the bacterium by forming a channel, called an infection thread, through which the bacterium penetrates the root hair and some of the underlying root cells. There the plant forms a welcoming chamber called a root nodule, where the plant supplies adequate nutrients for the bacterium to multiply, eventually forming a mass of bacterial cells which differentiate to become nongrowing cells called bacteroids. These bacteroids are nitrogen-fixing machines. They require oxygen, which the plant supplies in just the right amount for the bacteroid to make sufficient energy to fix nitrogen but not enough to inactivate its essential nitrogenase. The plant maintains this delicate balance by producing an oxygen-carrying protein quite similar to the oxygen-carrying protein hemoglobin in our blood. The plant's protein is called leghemoglobin. Like our hemoglobin, it is blood red.

Root nodules are readily visible on the roots of legumes. They have different distinctive, variable shapes and arrangements depending on the species of legume, but they all contain leghemoglobin. When crushed, all root nodules release a red liquid. The presence of the red color means that active nitrogen fixation is occurring, which is an easy and traditional way for a farmer to check the vigor of his or her legume crop.

The enormous capacity of prokaryotes to fix nitrogen would quickly deplete its atmospheric reservoir were it not constantly replenished. Prokaryotic microbes, both bacteria and archaea, are the only resuppliers of our atmosphere's essential reserve of nitrogen gas. They accomplish this task by either of two modes of metabolism: denitrification or anammox. Like nitrogen fixation, these are fundamental life-sustaining processes. In both of these processes nitrogen gas is a by-product of the way certain prokaryotes obtain metabolic energy from various forms of fixed nitrogen.

The relatively large number of prokaryotes that live by denitrification carry out a form of respiration in which they use nitrate ion (NO_3^-) instead of oxygen gas (O_2) to oxidize organic nutrients and thereby obtain metabolic energy. When oxygen is used, the end product is water; when nitrate is used, it can be nitrogen gas. In this form of respiration, nitrate is progressively reduced through a cascade of several reactions to nitrogen gas. Nitrite and nitrous oxide—otherwise known as "laughing gas," the mild sedative commonly used by dentists—are intermediates of the process. Some nitrous oxide, a powerful greenhouse gas, is released in the process. Denitrification has been studied in great detail. The process was discovered in 1882 by two Frenchmen, Ulysse Gayon and G. Dupetit. The driving interest of these studies was to find a way to suppress denitrification because it wasted fertilizer: much of the nitrates added to the soil by farmers was being converted to nitrogen gas along with some nitrous oxide gas by the denitrifying prokaryotes, principally bacteria, that are present in almost all soils. The key to controlling agricultural denitrification was found to be simple enough: keep the soil well aerated. Most denitrifying microbes are capable of using oxygen as well as nitrate for respiration, and because respiration at the expensive of oxygen yields more energy than when nitrate is used, these denitrifiers have evolved to use oxygen preferentially. Accordingly, soil aeration effectively suppresses denitrification.

The second process by which atmospheric nitrogen gas is replenished, anammox, wasn't discovered until a little more than a century later. In the 1990s, two Dutch microbiologists, Laura van Niftrik and Mike S. M. Jetten, were studying an anaerobic sewage digester. They noticed that ammonia was mysteriously disappearing. They found that the ammonia was being oxidized to nitrogen gas under these anaerobic conditions, hence the name anammox (*an*aerobic *amm*onia *ox*idation). Nitrite ion (NO_2^-), which was present in the digester along with certain then-unknown bacteria, was found to be the culprit. The nitrogen atoms in the nitrite ion and ammonia (NH_4^+) were combining and being released as nitrogen gas (N_2). The reaction releases a minuscule amount of energy, but it is sufficient to support the growth of several genera of these unusual bacteria. Subsequently, anammox has been shown to occur widely and profusely in nature. It has been estimated that over half of all nitrogen gas produced in oceans is a consequence of anammox.

Without microbes resupplying nitrogen gas via denitrification or anammox, the nitrogen cycle would be broken. Our atmosphere's nitrogen reservoir would soon be depleted, and the various forms of nitrogen on land would soon be converted, by the action of other microbes, to the highly soluble form nitrate, which would leach into the oceans and convert Earth's landmasses into nitrogen-depleted deserts. Although on a human scale this would seem to be a slow process, this catastrophe would happen almost instantly on the geological time scale because the turnover half-life of atmospheric nitrogen gas (the time required to replace half the gas) is a mere 20 million years.

Recently humans have intruded into the natural microbial monopoly of the nitrogen cycle. The consequences of these actions have been enormous: life saving, environmentally threatening, and for all practical purposes irreversible. Beginning near the middle of the twentieth century, humans became major participants in fixing nitrogen—converting atmospheric nitrogen to plant-utilizable forms. In the early twentieth century, an enigmatic German chemist, Fritz Haber, developed a process for reacting hydrogen (H_2) and nitrogen (N_2) gases at extremely high temperatures and pressures to form plant-utilizable ammonia (NH_3). Carl Bosch, a German chemical engineer, then industrialized the process. Later they developed another way to convert ammonia into the more traditional form of fertilizer, nitrate (NO_3^-). Their replication of two previously exclusively microbial processes, nitrogen fixation and aerobic ammonia oxidation, were at the time an astounding scientific and technological achievement. Gaseous nitrogen (N_2), with three bonds joining its two constituent atoms, is an extremely stable compound. That's why it demands so much energy from nitrogen-fixing prokaryotes. Ripping the two atoms apart so they can react with hydrogen to form ammonia presented a challenge that many engineers considered infeasible, particularly on an industrial scale. Myriad problems had to be solved. Thousands of catalysts were screened to find an effective one. The required temperatures and pressures are so high that the contained hydrogen reacts with the carbon in the steel of the reactors, weakening them and causing explosions. Bosch developed innovative, double-layered reactors to contain the extreme pressures.

The impact of the Haber-Bosch process is profound and pervasive. Today about half of all Earth's nitrogen, including the nitrogen in our bodies, was originally fixed by the Haber-Bosch process. In round numbers, each of us is about one and a half percent by weight a product of the Haber-Bosch process. Because of its monumental impact on agriculture, the Haber-Bosch process is sometimes deemed the most important invention of the twentieth century: it turned air into bread and it's the foundation of the Green Revolution, the comprehensive crop-improvement initiative spearheaded by Norman Borlaug, the American agronomist who in 1970 received the Nobel Peace Prize for his efforts. The Green Revolution increased wheat and rice yields and has allowed world agriculture to keep tentative pace with our burgeoning human population. For example, the Green Revolution converted Mexico from a wheat-importing to a wheat-exporting country and helped India produce enough to sustain its population. In India at least sixty million have starved to death in the past four centuries; in 1943 alone more than two million died during the Bengal Famine. The Green Revolution put a stop at least temporarily to these recurring tragedies. In 1966 India imported eleven million tons of grain; today, India produces more than 200 million tons.

Borlaug's approach was genetic: breed plants to produce higher yields. But Borlaug's improved varieties could yield more only because they were able to utilize more nitrogen, amounts that could be supplied only from the prodigious yields of the Haber-Bosch process. One of Borlaug's critical genetic innovations was the development of dwarf wheat. Its shorter, sturdier stems were necessary to support the heavy grain-bearing heads of the new, heavily fertilized varieties. To some of us, modern fields of wheat look like stocky, miniature parodies of the graceful, waving fields of our youth, but they produce a lot more wheat.

Today humans fix about as much nitrogen as microbes (bacteria and archaea) do, and we're locked into it. Without the Haber-Bosch process and its contribution to crop fertility, mass starvation would quickly ensue. Some credit the Green Revolution with saving as many as a billion human lives, but it has also contributed to environmental catastrophe. The dead zone is a coastal area the size of Connecticut and Rhode Island combined where the Mississippi River drains into the Gulf of Mexico. Excess nutrients (fertilizers containing nitrogen from the Haber-Bosch process as well as phosphate) run off from farms in the Midwest, drain into the Mississippi, and, dumping into the Gulf, cause eutrophication. The excess nutrients feed algal growth in the coastal waters, and these phytoplankton eventually sink to the ocean floor. Microbes decompose the phytoplankton, using available oxygen. The result is hypoxia, or oxygen-depleted waters that cannot support fish and other organisms that depend on oxygen.

Raoul Adamchak, an organic farmer in Northern California and past president of California Certified Farmers, mused, "Without nitrogen fertilizer to grow crops used to feed our recent ancestors so they could reproduce, many of us probably wouldn't be here today. It would have been a different planet, smaller, poorer, and more agrarian."

Use of Haber-Bosch derived nitrogen does not meet the strictures of organic agriculture and is opposed by many on that basis. Vandana Shiva, a charismatic and influential advocate of organic farming as well as an aggressive opponent of genetically modified organisms (GMOs), holds the Green Revolution responsible for the deaths of Indian farmers. She claims in her book *The Violence of the Green Revolution* that "Until the 1960s, India was successfully pursuing an agricultural development policy based on strengthening the ecological basis of agriculture and the self-reliance of peasants," a completely false assertion. But the influence of her emotional appeal to naturalism has been profound. Her influence has also successfully excluded from India the use of all genetically modified (GM) food crops, including golden rice, a vitamin-rich strain with proven health benefits. Richard J. Roberts, winner of the Nobel Prize in Physiology or Medicine in 1993, points out that millions of children have died or suffered developmental impairment because of lack of vitamin A in their diets and that use of golden rice could have reversed this. He posits that continued opposition to its use approaches a crime against humanity.

Although humans have commandeered about half of the fixation arc of the nitrogen cycle, its replenishment arc remains the exclusive province of microbes. Only they via the combination of denitrification and anammox can convert fixed nitrogen back to its gaseous form.

∽

Maintenance of atmospheric nitrogen is merely an example of our dependence on microbes. We wouldn't have evolved without them; we couldn't live now without their continued presence.

Our increasing environmental awareness and concerns, stimulated by environmental catastrophes and anticipated disasters, has engendered greater interest in microbial ecology. Many are taking a closer look at the critical roles microbes play naturally in keeping Earth habitable. We might make greater use of them to produce new and expanded sources of renewable energy and to clean up our waste—sewage, solid waste, oil spills, and toxic mining residues. As E. O. Wilson, perhaps our era's greatest biologist, put it when asked about what career he might choose today, "I would look to the major unexplored areas, at least for those with professional hopes. One of my favorites—one where I would go if I could start all over again now—is microbiology, particularly microbial ecology: the diversity of microorganisms, their ecology, and the study of the enormous impact they have on the planet."

In addition to the role microbes play in keeping Earth habitable, microbes may figure prominently in the future of human health. At the close of World War II, antibiotics promised an end to a massive scourge of bacterial diseases, from bubonic plague to tuberculosis. In

1967 William Stewart, then surgeon general of the United States, optimistically proclaimed, "The time has come to close the book on infectious disease. We have basically wiped out infection in the United States." Unfortunately, his statement was proved to be spectacularly wrong. In the interim many new bacterial infections have emerged, and the agents causing many of the old diseases have become resistant to antibiotics. Many of these antibiotics are now useless for battling certain infections. In the 1940s when the "wonder drug" penicillin was first available, *Staphylococcus aureus,* the causative agent of countless bacterial infections, systemic and local, was uniformly sensitive to the antibiotic, and the infections the bacterium caused were readily controllable by the drug. Doctors and veterinarians proceeded to use penicillin profligately and unwisely. I recall being given a mandatory prophylactic injection of penicillin in 1945 on leaving the USS *Cowpens* (CVL25). It was thought that sailors long at sea might have contracted gonorrhea, and a shot was necessary for liberty in San Diego. It was inevitable that resistant strains of *S. aureus* would develop. Now MRSA, methicillin-resistant *Staphylococcus aureus,* is dominant in hospitals and has become a major killer, responsible for over half of hospital-acquired infections. New, innovative ways to control infections by antibiotic-resistant bacteria are being pursued. There are some hopeful preliminary results.

Notable among these is the announced discovery in 2015 by Kim Lewis and colleagues at Northeastern University of a new antibiotic, christened teixobactin. The announcement is notable for several reasons. First, teixobactin is active against MRSA. Second, the researchers showed the way to tap a previously unexamined massive source of potential antibiotic-producing strains of microbes. An estimated 99 percent of soil-inhabiting microbes cannot be cultivated in the laboratory by conventional means. Using a device that the researchers have termed an iChip, they grew candidate microbes in pure culture in their native soil environment. The iChip is an array of small chambers, enclosed by a semi-permeable membrane. Single cells are placed in the chambers, and the device is buried in its native soil. Soil nutrients pass through the membrane to feed the microbe, which flourishes under these near-natural conditions.

The Northeastern researchers predict that MRSA and other microbes will not develop resistance to teixobactin owing to the unusual way it kills bacteria (a lethality restricted to Gram-positive bacteria). Rather than attacking one of the microbe's vital proteins, which can readily undergo mutation and confer resistance, teixobactin binds to and thereby sequesters two compounds (lipid II and lipid III) that are intermediates in the synthesis of the walls of Gram-positive bacteria. Lipid II is an intermediate in the synthesis of peptidoglycan, the major structural component of the bacterium's walls; lipid III is an intermediate in the synthesis of teichoic acid, a minor but vital wall component of Gram-positive bacteria. Deprived of these vital components, the cell's walls weaken and the cell bursts, driven by the force of its own osmotic pressure. Because teixobactin acts by binding to a vital cellular intermediate, which the researchers presume cannot be altered by mutation without dire consequences, the researchers predict bacteria cannot develop resistance to teixobactin. This strikes me as a plausible but overly optimistic conclusion that underestimates the power of natural selection. I remember a representative of Pfizer making the same claim for tetracycline, then the company's premier antibiotic. He was to a certain extent correct: the bacterial target of the drug didn't become genetically resistant in the conventional sense. Bacteria simply developed a method of pumping tetracycline out of the cell faster than it could enter. These mutant bacteria could grow in the presence of tetracycline; they became resistant, though not by a previously known mechanism. The more we know and understand about the Tree of Life and microbes' position in it, the more we are likely to be able to predict, or at least understand, microbes' behavior in the face of adversity.

The Tree of Life has thus far borne fascinating fruit, yielding more intimate information about our bond and debt to microbes. The tree has revealed how the cells of our bodies learned to function and how they acquired their complex metabolism, which allows them to utilize nutrients, grow, divide, and fulfill their varied essential roles that make our lives possible. Our cells accomplish many of their vital metabolic tasks in the same way that microbes do. When our cells and tissues take up a nutrient from our blood, they process it into cellular components, gaining metabolic energy as they proceed, through the same series of metabolic reactions that microbes invented and most continue to employ. When we run at top speed, our muscles use oxygen faster than it can be supplied. Our muscles then switch to an anoxic metabolism, attacking their reserves of glycogen and producing muscle-irritating lactic acid. To facilitate these various metabolic reactions, the enzymes our cells employ are structural homologues of those in microbes. The lactic acid in our aching muscles is made via the same set of energy-yielding reactions by which lactic acid bacteria cause milk to sour. We inherited these processes from microbes.

Is the Tree Complete?

Knowledge of the Tree of Life has allowed us to analyze the composition of the microbes that live in and on us. It's an impressive accomplishment, but it's not any more complete than a living tree. Nothing in science is beyond modification and growth as new information becomes available. Faith is immutable; science is alive, growing, and self-correcting. So far, all organisms discovered and examined can be rationally hung on the existing Tree of Life. There has been a place for all of them, but there will come a time where some change and reordering will be necessary.

The Tree of Life is, in fact, a scientific theory, an established summary and explanation of facts. Many of us define a theory as an opinion or untested proposal, but that doesn't apply in science. Even well-established scientific theories such as evolution are open to modification and growth when new and better information becomes available.

As new organisms are discovered, will the existing structure of the Tree of Life continue to accommodate all of them? Will new branches be added? These outcomes are impossible to predict. Only a little over thirty years ago, discovery of the third branch (the archaea) astounded biologists. It's entirely possible that representatives of a fourth branch might lurk in some of Earth's unexplored environments, perhaps in one of Antarctica's more than 140 subglacial lakes. The largest of these, Lake Vostoc, lies under about 2.2 miles of ice and is about the size of Lake Superior. Humans will also be probing elsewhere in our solar system. The venues for new life continue to expand.

Many life-packed ecosystems remain to be explored, and powerful new molecular methods are becoming available to test for the presence of microbes in them. Instead of having to culture, examine, and classify microbes in the ocean, for example, now it's necessary only to take the DNA from a quantity of water and examine it for the presence of unknown creatures. As we'll see, this new route to discovery is almost frighteningly productive, rapidly building our knowledge of interrelationships in the Tree of Life.

CHAPTER 2

Relationships among Organisms

The discovery of the archaea, a previously unrecognized domain of organisms, began with an attempt to understand the interrelatedness of various kinds of bacteria, a group of organisms with such simple structures that they defied the established methods of classification, which were founded on comparisons with morphology. New methods had to be developed. These were based on the structure of a cell's macromolecules rather than its visual attributes. This breakthrough led to the discovery that the Tree of Life has three distinct branches: the bacteria, the archaea, and everything else (the eukaryotes). This tripartite division is the fundamental basis of modern taxonomy.

Taxonomy

Taxonomy, like nomenclature, is a theory-neutral activity. We aggregate similar entities and name them without reference necessarily to the reasons for their similarity. These categories are as hypothesis-independent as the Dewey Decimal System. Classifying things is an ancient and compelling human activity, perhaps even an instinct. David Arora, the distinguished taxonomist of mushrooms, sometimes introduces his lectures by showing a photo of a group of toddlers sitting on a floor strewn with small wooden objects of different sizes, shapes, and colors. Some of the children are carefree, playing with one or several of the wooden pieces. But others are busily gathering them into groups of the same color, shape, or other eye-catching similarity without concern for why they're similar. The urge to classify must be in our genes. We express it early and often.

The history of biological classification, beginning with Aristotle, has largely mimicked the activities of the children in Arora's photo, substituting organisms for the children's wooden toys, of course. If organisms look alike, we group them together, name individuals, and then name the groups. This instinctive approach worked remarkably well for constructing coherent taxonomies of plants and animals, though it could not have happened without the aid of a rich fossil record that revealed the sequence of appearance and sometimes disappearance of certain organisms. Animals that had a backbone or red blood, for example, were grouped, as were plants that produced flowers or those that produced cones.

Almost by accident many of these initial similarity groupings proved also to reflect evolutionary or phylogenetic relatedness as well. More closely related organisms do tend to look alike. However, phylogeny was not the intent of the early taxonomists nor vital to their approach.

Carl von Linné, more commonly called Linnaeus, devised a scheme for classifying plants based on the anatomy of their sexual reproduction, the number and arrangement of stamens and pistils, which he felt reflected God's plan for plant life. He wrote in 1729 with more than a trace of anthropomorphism, "The flowers' leaves serve as bridal beds which the Creator has so gloriously arranged, adorned with such noble bed curtains, and perfumed with many soft scents that the bridegroom with his bride might there celebrate their nuptials with so much the greater solemnity."

Although today we might admire this eighteenth-century scientist's vivid imagination, a contemporary termed this approach to taxonomy "loathsome harlotry." Many of Linnaeus's reproduction-based groups of plants are indeed phylogenetically linked, but evolution was antithetical to his deeply held religious convictions. He believed that each species was a special act of creation and that it was immutable. As he wrote, "The invariability of species is the condition of order [in nature]."

It was this order, God's order, that Linnaeus thought a taxonomic scheme ought to reflect. Later in life, as a result of many confirming observations, Linnaeus did modify slightly his beliefs to entertain the possibility that new species might arise by hybridization between the *"primae specie,"* which he defined as those original species created by God that inhabited the Garden of Eden and named by Adam. He even entertained the possibility that new genera might similarly arise through hybridization.

Undoubtedly, Linnaeus's greatest contributions were those made to organismal nomenclature. He gave each species a Latinized binomial, a genus and species name (by convention, the former capitalized and both in italics, e.g., *Escherichia coli* or *Homo sapiens*). This brought a much-needed order to biology. Previously, scientists designated plants by short descriptive Latin phrases, which as we can easily imagine was awkward and led to taxonomic chaos. Linnaeus also instituted hierarchical groupings or ranking of species in ascending order into genera, families, orders, classes, and kingdoms, groupings that were later modified in the case of animals by adding phyla between kingdoms and classes. Aristotle had grouped species into genera; Linnaeus added the higher groups. His schema, a monument to logical nomenclature and taxonomy, remains largely in use today.

Linnaeus's belief that each species is an individual act of creation offered a crisp definition of the term and undoubtedly contributed to its nearly revered standing in biology. But as Linnaeus's definition of the word *species* became unacceptable, no satisfactory substitute emerged, in spite of the term being central to biological thought. Biology depends on the term, but can't define it. Darwin captured the dilemma in *Origin of Species:* "No one definition has satisfied all naturalists: yet every naturalist know vaguely what he means when he speaks of species." The dilemma has only deepened in the intervening years with our increased knowledge of microbes. Today probably the most widely used definition of species is the Biological Species Concept, offered by the great Ernst Mayr. Mayr defined species as "groups of actually or potentially interbreeding natural populations, which are reproductively isolated from other such groups." Of course such a definition has no relevance to the vast majority of living things, bacteria and archaea, that do not mate and hence do not breed in the conventional sense. Some biologists have turned to a substitute term, *phylotype,* but it, too, begs an acceptable definition.

The similarity-grouping system even brought useful order based on relatedness to our understanding of some microbes, notably the fungi. Those that produced similar structures, similar spores or arrangement of spores, for example, were grouped. The scheme was also usefully applied to the algae and for the most part to the protozoa, although later molecular-based studies tell a somewhat different story. Certain of these protozoan lookalikes prove to be only distantly related.

A Problem with Bacteria

Grouping organisms on the basis of their similar appearances, which proved problematic when applied to protozoa, largely collapsed when taxonomists turned their attention to bacteria. (Archaea, which offer similar taxonomic conundrums, hadn't yet been discovered to be a separate group, but the same dilemmas would apply to them.) There simply isn't enough recognizable structural diversity among bacteria to group them by appearance, let alone draw conclusions about relatedness within and among such groups.

Size is not very helpful. Some bacteria, principally ocean dwellers, are quite tiny, as small as a few tenths of one millionth of a meter (a micron, µ) in their longest dimension. A very few are truly giants, about a million times larger than the familiar *Escherichia coli*. Probably the smallest bacteria, at least known to us now, are those found in deep ground water at the Department of Energy Research site in Rifle, Colorado. One hundred and fifty of these miniature microbes could fit inside one *E. coli* cell. They have an average length of 0.3 microns, smaller than many viruses. Presumably their nutrient-poor environment exerts a powerful selective pressure for a greater surface to volume ratio and therefore the greater facility for uptake of nutrients that smallness confers. At the other extreme are the bacterial monsters. One of these is *Epulopiscium fishelsoni* (Fishelson's guest at a fish's banquet), which can grow to almost a millimeter in length (700 µm long and 80 µm in diameter). It is visible to the naked eye, thereby stretching the concept of microbe; it lives in intimate association with various species of the colorful, tropical surgeon fish. The other giant, about the same size as *E. fishelsoni*, is *Thiomargarita namibiensis*, which lives in the ocean sediments of the continental shelf of Namibia. Bacterial size therefore varies at least two thousand–fold. In spite of these examples, size does not offer a useful taxonomic metric. By far the majority of bacteria are between one and two micrometers.

The variety of shapes and colors of bacteria is also limited. Most are rods, some are spheres, and others are twisted like commas or corkscrews. There's not enough complexity or variation to make shape a useful taxonomic tool—we can't sort bacteria the way children sort blocks.

Bacteria do, however, differ profoundly by what they do. Indeed, the array of their diverse capacities to mediate biochemical transformations is astonishing. For example, collectively bacteria are able to utilize all naturally occurring organic compounds as nutrients and can make a living by doing so. Individually they differ enormously. Some can use petroleum and others cellulose, nutrients that no macrobe can attack. The variety of ways that various bacteria employ to obtain needed metabolic energy is also diverse and almost endless, in contrast to our limited abilities and those of our fellow eukaryotes.

We acquire metabolic energy only through aerobic respiration, that is, by employing oxygen to oxidize (metabolically burn) organic compounds as food, and the number of such compounds we can respire is quite limited. Not so with bacteria. Every naturally occurring organic compound is susceptible to being respired by some bacterium, a fact sometimes referred to as "microbial infallibility." Whether or not a particular bacterium can utilize a particular compound proved to be an excellent way to distinguish among bacteria, although it is not useful for assigning them to larger groups.

Although it is not useful for purposes of classification, what bacteria cannot do has become vitally important to humans. It's worth a brief digression from our journey to consider some of these lacunae. For the most part bacteria and other microbes are unable to utilize human-made plastics, which we continue to manufacture and which, because microbes can't degrade them, continue to accumulate in our environment, causing ugliness and overt environmental damage, choking birds and other creatures. Because most plastics float, oceans become repositories for huge quantities of them. Remote beaches, even if rarely visited by humans, have become desecrated. Some ocean-borne plastics never find their way back to land. Vast quantities of floating plastic debris have become trapped in giant cyclonic currents called gyres.

Worldwide there are five major gyres. The most notorious of them, the North Pacific Gyre, is more commonly called the "North Pacific Garbage Patch." It is notable for its extent (at least the size of Texas, possibly larger than the United States) and its abundance of garbage (about 5.5 kilograms of plastic per square kilometer). It may seem remote, but its ecological impact is immediate. Ninety percent of the Albatross chicks on Midway Atoll have been fed plastic, and 40 percent of them die from the plastic they have consumed. Consumed plastic can kill by blocking the intestinal tract. Lesser amounts can damage internal organs and confer a false sense of satiety, resulting in starvation. The US Fish and Wildlife Service calculates that the Albatrosses of Midway feed their chicks five tons of plastic a year. Albatrosses travel great distances to gather floating flying fish roe and other food for their chicks, but plastics in the North Pacific Garbage Patch are often taken home instead.

Why plastic is so invulnerable to microbes is not clear. Of course, plastic is relatively new, so microbes have had little time to evolve mechanisms to attack it. But that can't be the whole story. Some other human-made compounds, such as the explosive TNT (trinitrotoluene), are readily degraded by certain bacteria. One explanation is that plastics have extremely low solubility in water. The same is true of some pesticides, most notably DDT (estimated to be between 0.001 and 0.04 milligrams per liter). Nutrients enter prokaryotes and fungi, nature's major, nearly exclusive scavengers of chemical debris, only in soluble form; they have to be soluble in order to be metabolized. But the insolubility of plastics can't be the complete answer either. Cellulose, which a number of microbes, but no animals, readily degrade, is at least as insoluble in water as most plastics are. Microbes solved this metabolic challenge by secreting into their surroundings enzymes that break down the huge cellulose molecule into smaller, soluble fragments that the microbe can take up and utilize as nutrients.

Perhaps microbes merely need more time to evolve an ability to degrade plastics. It seems inconceivable that microbes will never become able to access this huge source of nutrition. It's a very welcome possibility that most microbiologists accept, but we shouldn't count on it happening in our lifetimes, if ever. A better approach would be to start using plastics that are currently biodegradable. Such plastics do exist. One of particular interest is a mixture of similar compounds called PHAs (poly-beta-hydroxyalkonates) that many bacteria in times of abundance accumulate as nutritional storage products, rather like we store fat, for subsequent use in leaner times. These biopolymers are commercially available but more expensive and thus not widely used. Legislation prohibiting use of conventional nonbiodegradable plastic bags may alter the equation. Many states and nations have already imposed or are considering bans.

PHA-based plastics were developed by Imperial Chemical Industries, a British company, and marketed in the United States by Monsanto, who, using recombinant DNA technology, engineered plants to make PHAs. Owing to the rising opposition to genetically modified organisms of all types, even those that would offer prodigious environmental and health benefits, Monsanto sold their rights to Metabolix, a company headquartered in Cambridge, Massachusetts. Metabolix offers a variety of PHA-based biodegradable plastics, including biopolymers for films and bags.

Perhaps the true costs, including their environmental insults, of nonbiodegradable plastics will one day be recognized, and use of those that are biodegradable will become widespread. The late William Jackson Payne of the University of Georgia proposed that biodegradability be the *sine qua non* for all synthetic compounds released into the environment. Even if a compound that is presumed safe and useful is released into the environment but later proves disastrous, microbes would rescue us from its long-term or permanent impact. If such a plan had been followed with the pesticide DDT, for example, the magnificent California Condor might no longer be threatened, almost forty years after DDT was banned in the United States for agricultural use.

Returning to what bacteria can do and their value to classification, bacteria are not only able to attack a vast variety of organic compounds, they attack them in a variety of ways in addition to oxygen-dependent respiration. Many can sequester metabolic energy from organic compounds in the absence of oxygen by such processes as fermentation (splitting a nutrient and using one part of it to oxidize the other) and anaerobic respiration (using a variety of oxidants, organic and inorganic, other than oxygen—sulfate ion [SO_4^-], for example). Some bacteria can capture energy from light via photosynthesis, and they can do this in more than one way, making them able to thrive in diversely lighted environments. The essence of photosynthesis is using light energy to reduce carbon dioxide to a cell-nourishing carbohydrate, so in addition to light, photosynthesizers need a source of reducing power. Plants and cyanobacteria use water for this purpose, casting off its oxygen as a waste product. Instead of water, some bacteria use reduced sulfur compounds such as hydrogen sulfide (H_2S); others use an organic compound.

Some bacteria don't even need organic compounds or light to prosper. They, like photosynthesizers, derive their carbon for making cellular constituents from atmospheric carbon dioxide rather than photosynthesis. Members of this class are called chemoautotrophs. They gain their metabolic energy by oxidizing inorganic compounds, such as certain reduced forms of iron or sulfur.

Some bacteria and archaea get their needed metabolic energy from chemical reactions that include highly unlikely nutrients. Methanogens, classified as archaea, are a good example of that. They obtain their metabolic energy by combining two gasses, carbon dioxide and hydrogen, to make a third gas, methane (CH_4); others can use solid iron oxide (rust) for the same purposes that we use oxygen as an oxidant for respiration. Other bacteria are able to exploit chemical reactions that yield tiny amounts of energy. Perhaps the record-holders are those that utilize the dab of energy released by stripping off one of the acid groups of malic acid found in grape juice (and elsewhere), thereby softening most red wines and rendering those from cold climates drinkable. Our partnership with bacteria has many facets.

Bacteria are also extremely disparate in their dietary needs. At one extreme, a major class of bacteria, the cyanobacteria, flourish on only what's available in air alone (carbon dioxide and nitrogen gas) if light is available, a feat no macrobe can accomplish without an assisting microbe. Plants, of course, also can live on carbon dioxide, but only because they harbor ancient bacterial symbionts, and those plants that are able to utilize atmospheric nitrogen depend on associated bacteria to do the job for them. At the other extreme, that of nutritional complexity, are certain bacteria, for example lactic acid bacteria, such as those we use to make cheese, yogurt, or sauerkraut. They are exceptionally demanding nutritionally. Some of them require a more diverse array of dietary supplements, amino acids, and vitamins in their diets than we do in ours.

So bacterial taxonomists elected to extend their classification database from mere observations on morphology to modest experimentation, determining what a particular bacterium could or couldn't do and searching for similarities among them. For example, was a bacterium able or unable to utilize a particular nutrient, perhaps the milk sugar, lactose, as a source of carbon atoms and metabolic energy?

Many such biochemically based tests were devised to distinguish among bacteria, and some were assigned a special significance. Ability to utilize lactose attained such a status. It became the primary means of distinguishing the bacterium *Escherichia coli,* most strains of which are benign, from its close, disease-causing relatives, including species of *Salmonella.* This test continued to be relied on as species defining in spite of the fact that a single genetic change, the kind that occurs frequently and spontaneously, can eliminate a bacterium's ability to utilize lactose. The resulting lactose-defective mutant strain can continue to flourish on other nutrients, but it appears to belong to another genus, according to this test.

Humans also differ among ourselves in our ability to metabolize this sugar. As infants, almost all of us can split it into its readily utilizable sugar components, glucose and galactose, but as we mature, many of us lose this ability to make the relevant splitting enzyme (Betagalactosidase); those of us who do become "lactose intolerant." Unable to digest it ourselves, lactose passes into our large intestines, where bacteria attack the sugar and produce acid and gas, causing intestinal distress. Those of us unable to consume ice cream, however, aren't likely to be placed in a different genus.

At one time it was extremely difficult to identify most bacteria in a natural environment. With modern molecular methods, now it's routine), but the pioneering bacterial taxonomist was forced to practice in the artificial and restrictive milieu of the laboratory. By using such admittedly artificial tests, taxonomists could isolate and identify particular bacteria, and of course they named them. Naming and identifying species is just the beginning of taxonomy, however.

Constructing a phylogenetic hierarchical scheme of classification for bacteria, one based on evolutionary relatedness, was until recently a seemingly unattainable goal for bacterial taxonomists or "bug sorters," as they were pejoratively dubbed by some. Identification of individual species of bacteria by biochemical testing was an important practical step forward for making microbiology a doable science, but determining how these various species are related to one another proved overwhelming. How could larger groups of relatives be recognized? How could individual bacterial species be stacked as logically as building blocks? Distinguished microbiologists despaired over this challenge. It became more a dream than a realistic research goal. How could groups of species be identified in the absence of recognizable, group-defining characteristics, such as possession of a backbone or the ability to flower? To bring some semblance of order to this vast, chaotic sea of individual names, microbiologists settled reluctantly on admittedly artificial schemes that might or might not have evolutionary support. Species were grouped by presumed linking characteristics. The logic of these hierarchical schemes was mere guesswork. As we'll see, some proved to be good guesses that did reflect evolutionary progression, that did have a phylogenetic basis. Others were not so good, and some now appear to be almost silly, such as speculations that bacteria evolved progressively to more complex shapes—that spherical bacteria (cocci) elongated to become rod-shaped, then twisted to become comma-shaped vibrios, and then continued to coil on themselves to eventually become helical-shaped spirilla. But there was a key to unlocking the conundrum of the actual phylogenetic relationships among bacteria, indeed among all living things, one just waiting to be found and exploited.

Macromolecules Hold the Key

Proteins—DNA and RNA—present in every cell provided the key to deciphering microbial relatedness. These huge molecules, macromolecules as they are appropriately termed, play disparate structural and metabolic roles but share certain common properties. They are all quite long (the molecule that comprises the chromosomal DNA of the bacterium *Escherichia coli,* for example, is about 1 mm long, some thousand times longer than the cell itself). Long, linked strings of various small-molecule building blocks—twenty different amino acids in the case of proteins and four different nucleobases of the nucleic acids, DNA and RNA—are joined together in a precise, role-defining sequence. The particular order in which these individual building blocks (monomers of the polymeric macromolecules) are lined up (their "sequence") varies from one individual protein or nucleic acid to another; it also encrypts the secrets of their evolutionary past.

In 1965 Linus Pauling, a chemist with a Nobel Prize for peace as well as chemistry—the only person to have been awarded two unshared Nobel Prizes—grasped the special status of these molecules as repositories of evolutionary history and christened them, accordingly, with an elegant and appropriate name: semantophoretic (from Greek, meaning "information bearing") molecules or semantides.

A few years earlier, in 1958, Francis Crick, the co-discoverer of the structure of DNA, is credited with grasping the same insight with his much-quoted comment that the amino acid sequence of a particular protein may be "the most delicate expression of the phenotype of the organism, and that the amounts of evolutionary information may be hidden away within them." Pauling, however, stated the hypothesis explicitly and offered experimental proof of its validity.

Pauling came close to adding another major achievement to his long list, namely deciphering the structure of DNA, a puzzle that he was exceptionally equipped to solve and with which he was already actively engaged. Pauling lacked some critical information: the photos of X-ray diffraction patterns of DNA prepared by Rosalind Franklin in Cambridge. Pauling had been invited to attend and speak at a scientific conference in London in 1952, but his passport was revoked for political reasons. He had joined the Emergency Committee of Atomic Scientists chaired by Albert Einstein, an organization dedicated to informing the public of the dangers associated with developing nuclear weapons. Such membership was judged subversive by the US government, eager to develop its nuclear weapons program at the outset of the Cold War. Thus Pauling missed any opportunity he might have had to meet with Franklin.

Franklin was working at King's College, Cambridge with Maurice Wilkins and with the aid of a graduate student, Raymond Gosling. Wilkins, without Franklin's consent, showed Gosling's best image, image 51, to James Watson. The image clearly indicated that DNA has a helical form. Watson and Crick used that information to deduce DNA's structure, for which Watson, Crick, and Wilkins, but not Franklin, shared the Nobel Prize. Had Pauling seen that DNA diffraction image, the story of the determination of DNA's structure could have been quite different.

The special status of the molecules that Pauling termed semantides derives from their being the molecules of which genes are made or molecules that carry the information contained in genes. Of all the types of molecules found in cells, only three, all macromolecules—DNA, RNA, and protein—qualify as semantides. Lipids, carbohydrates, and all the cell's other macromolecules such as lignin or peptidoglycan don't qualify. These other macromolecules are genetically determined, of course, as are all the cell constituents, but they are not direct carriers of genetic information. They are not copied from DNA or a molecule that has been copied from it. Instead these other macromolecules are made via reactions catalyzed by enzymes, proteins with the power of facilitating reactions (known as catalysts). Like all proteins, enzymes also contain genetic information. And so macromolecules that are not semantides are on a dead-end informational sidetrack, isolated from the direct stream of genetic information that flows from DNA to RNA to protein.

The order of the constituent building blocks that make up semantides precisely records evolutionary history. These building blocks are amino acids in proteins, and nucleobases both in DNA (adenine, guanine, cytosine and thymine) and in RNA (adenine, guanine, cytosine and uracil). Every protein in every cell contains a huge reservoir of information that reveals its own evolutionary history and often that of the cell in which it resides. Consider, for example, the information content of hemoglobin, the red, oxygen-carrying protein found in the blood of mammals. Hemoglobin is composed of strings of 287 amino acids, each of which could be one of the twenty different kinds of amino acids, about the same number as letters in the English language. Imagine, as Scrabble players know, the almost unlimited number of words and their implied thoughts, ideas, or concepts that could be expressed by arranging 287 letters in various sequences.

Using the structure of semantides to trace evolution was a great intellectual achievement. Even so, its fundamental premise is straightforward, and like most great intellectual breakthroughs, even obvious after the fact. The premise depends on just a few facts and suppositions. First, as we've noted, particular proteins are composed of a long string of amino acids arranged in a specific sequence. (The same applies to the string of nucleobases that make up their encoding DNA and RNA.) Second, the sequence of amino acids in a particular *type* of protein, hemoglobin, for example, that fulfills the same function in different organisms is similar in each of them but not quite identical. Such differences presumably arose stepwise as evolution proceeded from a common ancestral protein in a common ancestral organism. These differences among organisms are the consequences of mutations that occur randomly but inexorably over evolutionary time. The overwhelming majority of such changes, like all mutations, are deleterious and so would have been eliminated by natural selection. We wouldn't see them retained in extant organisms. But a minority would have been neutral (causing no detectable change), tolerated (causing minimal change), or preferential (bettering the protein's activity at least for the particular organism that houses it). These various mutations would inevitably accumulate over time.

So the number of differences in the sequences of amino acids in the same type of protein present in two different organisms must reflect the evolutionary time separating them. In this way we can judge the relative evolutionary distance from a common ancestor and thus from each other, or can we? Mutational change is random and in most cases genetically reversible. A subsequent mutation can return a prior mutation to its original state, thereby apparently minimizing the actual evolutionary distance. To estimate the number of changes that must have occurred to generate the observed differences between two semantides, tree builders resort to statistical probabilities, using one of several schemes. The sequences of nucleobases in nucleic acids tell the same story—an all encompassing one capable of revealing evolutionary history and the Tree of Life. Not all organisms have hemoglobin, but they all contain DNA and certain RNAs.

Of course, there are limits to the number and kinds of mutational changes that a particular type of protein can tolerate before it loses its identity. If hemoglobin, for example, loses its ability to carry oxygen, it then ceases to be hemoglobin. Natural selection prohibits these sorts of changes as well as what the microbial geneticist John Roth has called "purifying selection," the elimination through competition of mutations that cause a protein to function less well. But proteins have proven to be extraordinarily pliable nonetheless. Most tolerate many changes before they are damaged or lose their functional identity, and a sufficient number of these reveal a meandering trail of evolutionary history.

In 1965 Linus Pauling put to the test the idea that sequences of amino acids in proteins were the Rosetta Stone of evolution. He and his colleague Emile Zuckerkandl stated the premise clearly and tested its general utility for revealing relationships among animals. To evaluate its practical value, they chose to study hemoglobin, which plays the same critical role in all organisms that contain it, namely, binding to oxygen from inhaled air and carrying that oxygen to the body's tissues. Curiously, some two millennia earlier, Aristotle had unknowingly assigned great taxonomic significance to the same protein by dividing animals into two groups: those with and without red blood (i.e., containing or lacking hemoglobin), which as it turns out is roughly equivalent to having or lacking a backbone.

Pauling set out to test his hypothesis by determining how the differences of the sequences of amino acid in hemoglobins among a group of animals correspond to their known relatedness. His challenge was considerable. At the time, determining the sequence of amino acids that make up a protein was a daunting chemical undertaking. Each amino acid in a protein's chain must be chemically removed and identified, from the last to the first, each requiring a time-consuming chemical procedure. Hemoglobin, a moderate-sized protein, contains 287 component amino acids made up of two chains, one with 141 constituent amino acids and the other with 146. Determining and comparing the sequences of amino acids in a set of such proteins, though theoretically possible, would have been impractical.

So Pauling followed an innovative and, as it developed, trailblazing alternative route. In doing so, he relied on another biological treasure: the precision with which enzymes act. Most vital cellular reactions occur at a measurable rate only when they are enabled (catalyzed) by a particular enzyme, so organisms must make large numbers of different enzymes in order to catalyze each of their essential chemical reactions. *E. coli,* for example, makes around a thousand different enzymes. A few enzymes catalyze a broad set of similar reactions. For example, enzymes called lipases catalyze the degradation of a wide variety of fats and oils. But most other enzymes are precise and discriminatory with respect to the reactions they enable. One enzyme, D-lactate dehydrogenase, acts on only one form of one compound, D-lactic acid. It cannot act on even its mirror-image form,

L-lactic acid, though the compounds are as similar to one another as our two hands.

Pauling took advantage of the remarkable specificity of certain enzymes called proteases, which fragment proteins into smaller pieces. An organism's various proteases play disparate roles in its metabolism, but notable among them is digestion, the breaking down of proteins into more usable pieces by cleaving certain of the chemical bonds that link a protein's constituent amino acids. In doing so, proteases cut long molecules into a number of fragments. Proteases do not act randomly, a fact vital to Pauling's investigation. Rather, they cut at specific locations and only between particular pairs of amino acids in the chain. In a sense these proteases read the sequence of amino acids that make up a protein and then cut it at their preferred location. So, the number of cuts that a particular protease makes in a protein and the number and composition of the resulting fragments yields information about the sequence of the amino acids that make up its chain. Thus proteases have a capacity, although somewhat limited, to yield information about the sequence of amino acids that compose a protein. Pauling recognized this ability and exploited it.

For his studies on hemoglobin, Linus Pauling chose the protease called trypsin, an enzyme found in the duodenum of many vertebrates, including our own, where it aids our ability to digest the proteins we ingest. The rules governing where trypsin cuts proteins are precise and somewhat arcane: it cuts the amino acid chains of proteins only on one particular side of either of two amino acids in the string, lysine and arginine (unless they are followed on the cut side by the amino acid proline). So if a lysine or an arginine residue in the protein chain were, through the course of evolution, changed to some other amino acid, trypsin would no longer cut there; the consequence would be fewer cuts, and thus fewer but larger resulting protein fragments. Adding one of these amino acids or eliminating certain prolines increases the number of cuts and resulting fragments. As a consequence, the size of the fragments yielded by trypsin cutting is a measure of how many amino acid units there are and where they lie in the chain, a crude reflection of the sequence of amino acids composing the chain. Pauling hoped that the order of the chain would reflect its evolutionary history.

Pauling cut hemoglobins from twelve quite different animals with trypsin and examined the number, chemical properties, and sizes of the resulting fragments. Trypsin severed most of these hemoglobins into about two dozen fragments. The results told a very encouraging story and answered some important questions. First, the pattern of fragments from each of the hemoglobins was different. That reassured Pauling that hemoglobin and presumably other macromolecules as well do change substantially over immense evolutionary time but still retain their identity and function, an encouraging finding for the future of the semantide approach to tracing evolution, offering hope that perhaps a complete history of evolution could be recorded in a single macromolecule, were the right one to be identified.

Even more important, the different numbers and sizes of fragments in patterns Pauling found correlated with our knowledge of the evolutionary distances separating the animals from which the hemoglobins were obtained. The fragments from the hemoglobins of closely related animals were more similar in size, number, and properties than those from more distantly related animals. For example, there were fewer differences between the hemoglobin fragments from chimpanzees and humans than between those from orangutans or other primates and humans, indicating, as we now know to be the case, that humans are more closely related to chimpanzees than to orangutans. Of all species examined, the hemoglobin fragments from the lungfish differed the most from those of human hemoglobin, making the lungfish our most distant relation.

In 1967 Allan Wilson, a biochemist, and Vincent Sarich, an anthropologist, confirmed the relationship-exploring value of semantides and discovered that semantides could also be used to estimate the time that evolutionary events took place. In Wilson's words, semantides can function as a "molecular evolutionary clock," encoding information about when the Tree of Life branched. Before his death from leukemia at age fifty-six, Wilson established himself as a leader in applying molecular analyses to studies on evolution. He was also a much-admired teacher and a good scientific citizen. When Carl Woese was unable to meet his obligation to present the Roger Y. Stanier Memorial Lecture at Berkeley, Wilson gave a brilliant lecture with less than a full day's notice.

Wilson's approach to exploiting evolutionary history encrypted in semantides was somewhat different from Pauling's. He chose a different protein, namely one called serum albumin, which is abundant in the blood of most animals, and a different way of studying its molecular sequence. He exploited the discriminatory powers of immune reactions using a more restricted range of organisms—primates.

Antibodies, the immune system's powerful front-line weapon against invading microbes, form in response to exposed proteins on the microbe's surface; in subsequent encounters with the same microbe they inactivate it by binding to that exposed protein, thereby conferring immunity against a subsequent attack by the same microbe. The binding capacity of antibodies is even more precise than that of enzymes. Antibodies form in response to a purified protein and react with that particular protein, but they also "cross-react" less vigorously with very closely related proteins. Wilson reasoned that the degree of cross-reactivity between two proteins, which can be quantified, must be a measure of the differences of their sequence of amino acids and therefore their relatedness. Wilson measured the level of cross-reactivity of serum albumins from various primates using a quantitative technique called complement fixation. He injected rabbits with purified serum albumins from various primates and tested reactivity of the resulting antibodies they formed against other purified serum albumins, giving him a numerical index that he dubbed the index of dissimilarity (ID).

These indices largely agreed with what was already known about relationships among primates: our closest primate relatives are chimpanzees (ID 1.14), bonobos (ID 1.40), and gorillas (ID 1.09); we are more distantly related to orangutans (ID 1.22), gibbons (1.28), and siamangs (ID 1.30). A more surprising result showed that chimpanzees are more closely related to us (ID 1.09) than they are to gorillas (ID 1.17). Less surprisingly, by this index, chimpanzees are indistinguishable from bonobos (ID 1.00).

Wilson then turned his attention towards a new question: Can the rate of evolutionary change in proteins serve as an evolutionary clock, at least within the primates? Is the ticking of the occurrence of sequence-altering mutations sufficiently constant? At first blush this seems spectacularly unlikely because mutations occur randomly, as randomly perhaps as the decay of particular radioactive atoms. But randomness does offer its own statistically collective regularity, if sufficient events are examined. For example, rate of radioactive decay is the basis for our most precise measurements of time.

First Wilson asked whether the genetic clock ticks at a constant rate. If it does, he reasoned that a set of closely related primates that descended from common ancestors must have the same index of dissimilarity (ID) from those ancestors, because they had followed equidistant evolutionary trails. In his test he chose representatives of Ceboidea (New World monkeys), Hominoidea (apes and man), and Cercopithecoidea (Old World monkeys) as the set of related primates. These had all descended from the same distantly related set of primates, the Prosimii, which includes lemurs, lorises, bush babies,

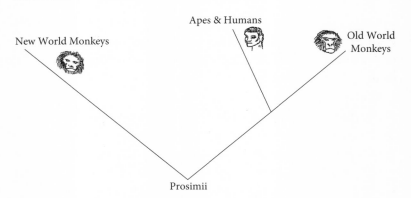

FIGURE 7. Scheme by which Allan Wilson calibrated the evolutionary clock of primates. New World monkeys, apes, and humans, as well as Old World monkeys, as shown all evolved from prosimii by different but equidistant routes. If the evolutionary clock for primates runs at a constant rate, each of these groups of monkeys and apes should have the same index of dissimilarity to prosimii, which proved to be the case.

and tarsiers. The evolutionary distance from all the closely related primates to the more distant branch point must be the same. But were their protein differences the same? That proved to be the case. The protein differences, ID, between representative Old World monkeys, New World monkeys, and *Homo sapiens* from Wilson's chosen example of Prosimii, the galago, was 9.0, 11.3, and 10.8, respectively, all clustering closely enough to indicate that the evolutionary clock, at least in this case, ran at essentially the same speed in these three distinct lines of descent.

Wilson then calibrated his clock. He did so on the basis of traditional means of reckoning that the Old World monkey line branched 30 million years ago from that leading to other primates closely related to humans. With such a calibration, Wilson's ID values then showed that the line of descent leading to humans and that leading to the New World monkeys branched only 10 million years ago. Wilson's clock further indicated that we share a common ancestor with our very close cousins, chimpanzees and gorillas, and our somewhat more distant cousins, the orangutans, lived only 5 million years ago, a split second of evolutionary history.

However, the evolutionary clock that runs so reliably within the primates speeds up and slows down dramatically within other lines of evolutionary descent. Although this result is disappointing, we shouldn't be surprised. Rates at which mutations occur and means of correcting them differ dramatically among organisms. In the case of some microbes, these mutations and corrections can happen even between some that are closely related. For example, *Oenococcus oeni* is the bacterium that carries out malolactic fermentation, which converts a wine's dicarboxylic malic acid to the less acidic monocarboxylic lactic acid. *O. Oeni* corrects mutations much less effectively than its close cousins who play more mundane roles, such as converting cabbage into sauerkraut. Thus the semantides of the bacterium that mediates the malolactic fermentation evolve more rapidly.

∼

Pauling's result with hemoglobin provided no important new information about relatedness, and Wilson's with serum albumin contributed only a little more, but they did offer a new approach for answering biology's most fundamental questions about the relationships among living things. By identifying the fundamental biochemical similarities that all organisms share, we are now able to state with some certainty that all cellular organisms are related; they are made of the same kinds of macromolecules—proteins, nucleic acids, polysaccharides, and lipids. (The noncellular entities we call viruses are excluded from this definition. Whether or not one considers them to be alive, a popular but unfruitful discussion, their origin and evolution remain a mystery.) And with very few exceptions, these macromolecules are built from the same component parts.

Proteins from all organisms contain some or all of the same set of twenty amino acids. (A minority contains a twenty-first.) There are different numbers of amino acids in different proteins. They are arranged in different orders, but their building-block amino acids are always the same.

DNA from all organisms, which differs so profoundly in size and sequence, is built from a common set of nucleobases: adenine (A), guanine (G), cytosine (C), and thymine (T).

RNA follows the same pattern of overall difference but component similarity: RNAs from all organisms are built of adenine (A), guanine (G), cytosine (C), and uracil (U) instead of DNA's thymine.

Possibly the most striking similarity among organisms is their common use of the same genetic code: the instructions for making proteins are written the same way in all organisms that have been examined. For example, as we'll see, a T followed by an A and then a C in the sequence of DNA that makes up a particular gene mandates the cell that bears it to insert a molecule of the amino acid tyrosine into the protein that is being made according to that gene's instructions. Regardless of whether the cell is microbial, plant, or animal, this instruction is always followed the same way.

Fundamental challenges remain in unravelling the entire Tree of Life. Pauling's approach depended on studying certain proteins, those such as hemoglobin with similar functions in the organisms in which they are found. And hemoglobin is found only in a fraction of living things—just red-blooded mammals. Studying hemoglobin can tell us about our relationship to animals with backbones but not to other animals, plants, or microbes. Serum albumin can tell us only about relationships among primates. Clearly, a universally distributed semantide is needed to better understand all the branches of the Tree of Life.

Many proteins are more widely distributed than hemoglobin. For example, the respiratory protein cytochrome c is present in almost all organisms and has been recently studied as a potential tracer for relatedness. In 1967 Walter Fitch and Emmanuel Margoliash made a particularly careful analysis of the relatedness of twenty organisms using published information about the sequence of amino acids in their cytochrome c. All were eukaryotes, although three were microbes—two yeasts (*Saccharomyces* and *Candida*) and a filamentous fungus *(Neurospora)*. They compared these proteins by noting changes of amino acids at particular locations. Then they tallied the number of base changes that would have had to occur to encode these transitions from one amino acid to another, thereby giving them a numerical index of differences between pairs of organisms. With these numbers they constructed a tree of relatedness, which closely matched what was already known from biological observations. For example, donkeys were closely matched to horses, and monkeys to humans. The microbes were the most distantly related to humans. Once again these studies confirmed the validity of Pauling's hypothesis: semantides encode the history of life's evolutionary history.

In spite of the richness that Fitch and Margoliash found in the story told by cytochrome c, this protein is not universal, either, and is likewise incapable of showing us a complete set of relationships within the Tree of Life.

Ribosomes

There is a vanishingly small set of semantides that could fulfill the requirements necessary to unveil the entire Tree of Life. First, the right one must be present and serve the same function in all living things.

Furthermore, a tree-revealing semantide must be sufficiently tolerant of changes in the sequence of its component building blocks: fewer differences would indicate closer relationships, and a large number of differences should indicate the relationship between two very different creatures, such as the bacterium *Escherichia coli* and *Homo sapiens*. Such a semantide does exist. It's not just a protein. It's a large, composite structure of proteins and RNA called the ribosome. The complexity of its composition is not a roadblock. The sequence of nucleobases in molecules of RNA can encode the same evolutionary history as the sequence of amino acids in proteins. Like protein and DNA, RNA is a semantide.

If the sequence of monomers making up a semantide is the script of biology's evolutionary Rosetta Stone, ribosomes are certainly the slab upon which the information is inscribed. Ribosomes are the biologically ubiquitous, information-packed, intracellular particles where proteins are synthesized. All cells make protein; cells of all organisms contain ribosomes. Some cells are packed with them. For example, each cell of the well-studied bacterium *Escherichia coli*, when growing rapidly in a favorable environment and therefore rapidly synthesizing proteins, contains about 70,000 individual ribosomes. In such a cell ribosomes make up about half (45 percent) of the cell's total mass. They are the structures that give the cytoplasm of prokaryotes its grainy appearance under the electron microscope. Ribosomes are expensive for a cell to make, so most bacteria and probably other cells as well make only as many ribosomes as they need. When growing in a nutritionally poor medium that supports only a slow rate of growth, they make only a few ribosomes. Otherwise the cost of making extra ribosomes would cause them to grow even more slowly. When growing in a rich medium that supports rapid growth, such cells pack themselves with ribosomes.

The discovery of these essential cellular tools of protein synthesis is surprisingly recent, within the span of my own scientific career. I met them in graduate school, before they were christened and before anyone knew what they are and what they do. Roger Y. Stanier introduced them to me as a scummy layer on the bottom of a centrifuge tube, the jetsam of an experiment in which he was engaged.

Stanier was curious to know the form of the RNA and protein that remained in cell extracts after the microscopically visible cellular structures had been removed by low-speed centrifugation (about 7,000 rpm). He had engaged the help of two young physical chemists, Howard Schachman and Arthur Pardee, who had access to an analytical ultracentrifuge, then a somewhat expensive and not widely available instrument. Stanier was in distinguished company with both the instrument and his collaborators. The instrument, then state of the art though now replaced by more precise machines, was the brainchild of the Nobel Prize–winning Swedish chemist Theodor Svedberg. With it Svedberg had already discovered two fundamental properties of proteins—their molecular weight and their homogeneity. He learned how huge proteins can be, and that each molecule of a particular kind of protein, hemoglobin for example, has the same molecular weight. One of Stanier's collaborators, Howard Schachman, had already made important contributions to the development of the ultracentrifuge, and he would go on to make more. The other, Arthur Pardee, would become a major star in the burgeoning field of molecular biology, participating in the legendary PaJaMo (Arthur **Pa**rdee, François **Ja**cob, Jacques **Mo**nod) experiment that laid the foundation for understanding how the synthesis of many enzymes is regulated.

An ultracentrifuge rotates at speeds high enough to subject samples to intense centrifugal forces around 130,000 times greater than the force of gravity, powerful enough to cause particles such as ribosomes to migrate through the column of water in a spinning centrifuge tube. Schachman and Pardee's ultracentrifuge was also able to characterize the size and shape of small particles to a certain extent because it was an analytical centrifuge, outfitted with an optical device that is capable of following the rate of migration of particles subjected to these forces. The rate of such migration is measured in "S value" units, named after the device's inventor, Svedberg, who had developed the theoretical basis of the measurement, an index of the mass and shape of small objects. The higher the mass and the greater its streamlining shape, the faster a particle migrates in response to centrifugal force and the greater its S value.

Stanier proposed that his colleagues use their ultracentrifuge to answer a simple question about the internal structure of prokaryotic cells: What's left in their cytoplasm after all the microscopically visible particles have been removed by low-speed centrifugation? Stanier already knew from chemical analyses that RNA remained. He wondered about its structure. Did this RNA exist in the form of small particles or was it in solution? Something did indeed sediment in the ultracentrifuge, forming a scum at the bottom like the one that Stanier showed me. The conclusion was clear. Some of the RNA and protein is present in the form of particles. This finding by itself was not particularly surprising or interesting. To make an extract of bacteria, that is, to obtain a sample of their cytoplasmic interior, cells have to be ruptured, which requires drastic treatment because the walls of most bacterial cells are extremely tough. They can, however, be broken by a variety of harsh treatments: by intense sonic fields, grinding with abrasives such as alumina, or by subjecting them to an intense hydrostatic pressure and suddenly releasing it. Stanier was concerned that the friendly-fire trauma of breaking the cells might have torn fragments from cellular structures and that the particles that settled in the ultracentrifuge were artifactual trash, merely the resulting nondescript cellular debris.

But the S-value-measuring optical device on the analytical ultracentrifuge showed that this was not the case. The extracts were not filled with a complex mixture of particles of various sizes and shapes, as one would expect to see in the debris formed from harsh mechanical treatments. Rather, they contained a homogenous population of just a few kinds of well-defined particles of particular sizes (that we now recognize as being the several subunit components of ribosomes). The ultracentrifuge saw these same particles, regardless of the way he ruptured the cells. So they weren't produced as artifacts of the destruction of the cell. They were discrete, pre-existing intracellular structures.

Stanier wondered if all bacteria contain such particles. Accordingly he looked for them in seven other species of bacteria. All contained these small structures, and they were the same size in each of the seven species. Finding these structures in seven bacterial species doesn't prove that all bacteria contain them, but it's powerfully suggestive. What about all organisms? He looked at cells of a eukaryote, baker's yeast *(Saccharomyces cerevisiae)*. They contained quite similar particles, only a little larger. (We now know that ribosomes in eukaryotes are just a bit larger than those in the other two biological domains, the bacteria and archaea.)

Stanier was fully aware that these "microsomes" (small bodies), as he called them, the components of the scum on the bottom of the tube that he showed me, must be doing something essential. Why else would all these cells contain such large numbers of these highly uniform structures? But he had no hints as to what their cellular function might be.

Soon, because of their ubiquity and abundance, these intriguing particles attracted wide scientific interest. Chemical analyses showed that they were composed of about equal parts of protein and RNA, making them even more intriguing. At that time, in the early 1950s, the function of RNA was unknown. Was it a precursor of DNA? Did it play some other vital but unknown role?

Shortly thereafter these "microsomes" got a proper name of their own. In 1958 at a symposium to discuss these still mysterious ribonucleoprotein particles, Richard B. Roberts, a polymath who had made important contributions to nuclear physics, weapons technology, biochemistry, molecular biology, and neurobiology, proposed that they be called "ribosomes." The term seemed to describe them well, implying structures or bodies composed of RNA, and moreover, as Roberts said, "It has a pleasant sound." His suggestion stuck.

But what is the cellular purpose of ribosomes? The first experiments revealing what ribosomes actually do were made in 1954 by Paul C. Zamecnik, who would make many important contributions to biochemistry and medicine, along with E. B. Keller. They added radioactively labeled amino acids to animal tissues that had been ground into a paste. This mixture of cellular debris retained the intact cell's capacity to string together added amino acids, incorporating the added radioactive amino acid to produce radioactive proteins, an important discovery in itself. More importantly, they discovered that the newly made radioactive proteins were at first attached to ribosomes. Only later did they detach and become free in the mixture.

Zamecnik and Keller concluded, correctly, that amino acids had become linked together on the ribosomes. Later, when a complete protein had been assembled, it was released from its ribosomal birthplace. Their experiment showed that ribosomes are the platforms or the workbenches where proteins are synthesized. All organisms make their own component proteins. That's why all organisms contain ribosomes; they must have them to grow.

Not only are ribosomes biologically universal, they are remarkably similar in all organisms. Throughout their long evolutionary history they have not changed in any fundamental way, but have continuously undergone minute changes, dropping molecular crumbs that leave behind a distinct, traceable evolutionary trail.

All ribosomes are composed of three RNA molecules of different sizes: small, medium, and large, as well as over fifty different protein molecules. There are some variations of the overall molecular structure of ribosomes among organisms, particularly with respect to the sizes of their RNA components. These differences break down along dominion lines. The ribosomes of bacteria and archaea (70S) are somewhat smaller than those of eukaryotes (80S). Still, as we'll see, the similarity of sequences of the nucleobases in these RNA molecules clearly shows they are related and share a common ancestor, presumably the universal ancestor of life.

Certainly ribosomes meet the primary requirement for translating evolutionary history and therefore relatedness: all living things contain them. But ribosomes are not a single macromolecule; they're a collection of them. As we've noted, each ribosome contains three different molecules of RNA and fifty to eighty different proteins in various organisms. Which of these molecular components best records the core information of evolutionary history? Ribosomes must have existed throughout the history of cellular organisms, but did all of their components? Did any of the ribosome's encoding genes witness the evolution of cells from their beginning, and can they tell us about that history?

At about the same time these questions were being debated, a major advance in biology provided many answers, both about ribosomes and about our ability to describe the Tree of Life.

CHAPTER 3

Enter DNA

In 1953 a brash twenty-five-year-old American, James Watson, and his collaborator, an innovative thirty-six-year-old Englishman, Francis Crick, published in *Nature,* the prestigious British journal of general science, a two-page article with the modest title, "Molecular Structure of Nucleic Acids: A Structure for Deoxyribose Nucleic Acid." Although studies on the molecular structure of biomolecules often make solid and lasting contributions to the advancement of biology, they rarely attract immediate wide interest. Not so with this paper. Its impact was instant and far reaching. I recall my boss at the time, Tom Wood (who first isolated vitamin B12), bouncing into my lab with a copy of the journal in hand. He had to share his excitement with someone. He sensed that the paper would change biology profoundly. He was certainly correct.

The Watson and Crick paper presents more than the structure of an important biological macromolecule. It rebuilt the foundations of biology and human affairs by suggesting how inheritance works. By explaining the structure, they describe how the DNA double-helical molecule might encode genetic information and how it could be copied accurately so that a cell's cumulative instructions for life could be passed on to its progeny. Watson and Crick wrote in the final, understated sentence of their article, "It has not escaped our notice that the specific pairing we have postulated immediately suggests a possible copying mechanism for the genetic material." That "specific pairing" provides the answer to one of biology's most elusive questions with respect to the unique nature of genes. J. B. S. Haldane put it succinctly some sixteen years previously: "one essential property of the gene is that it reproduces its like at each nuclear division." DNA's structure showed how that could happen.

The "specific pairing" Watson and Crick refer to is represented by the chemical bonds that hold DNA's two component single strands together, forming the now legendary double helix. The two single strands are long strings of alternating sugar (deoxyribose) and phosphate molecules with a nucleobase, either guanine (G), cytosine (C), adenine (A), or thymine (T), attached to each phosphate group on each of the two strands. The nucleobases extend toward the molecule's central core, holding the two helical strands together by the specific pairings that Watson and Crick emphasize. Watson and Crick show that these pairings are the relatively weak chemical bonds called hydrogen bonds, bonds formed by sharing hydrogen atoms rather than electrons, as the stronger, more familiar covalent chemical bonds do. Because the sharing atoms are in just the right position, such hydrogen bonds form spontaneously between specific pairs of nucleobases: G attaches to C by three hydrogen bonds, and A attaches to T somewhat less firmly by two such bonds. This specific pairing is the essence of what DNA is—how it directs cellular activities and records and maintains evolution's legacy.

Such specific pairing suggests how life's inheritance is preserved and how a double helix is precisely replicated. First, enzymes split the double helix into its two constituent single strands. Building blocks (called nucleoside triphosphates) of a new strand of DNA line up by forming hydrogen bonds between the particular nucleobase they contain and the complementary nucleobase jutting from the single strand. Then another enzyme links these building blocks together, forming two double helices. Each of these contains a sequence of nucleobase pairs identical to those in the original (parent) double helix, and each contains one newly made strand.

At the time of Watson and Crick's article, it had recently been established that genetic information is stored in DNA, encrypted in the sequence of its constituent nucleobases. Their breakthrough suggested that DNA must be the repository of evolutionary history as well.

In 1958 in a lecture for the Society of Experimental Biology, Crick proposed what he called biology's central dogma: the information stored in DNA is shared via RNA with proteins, but proteins cannot send information back to DNA. Pauling might have called this the informational pathway of the semantides. As Crick put it, "Once information has passed into protein, it cannot get out again." Because they are links in the unidirectional freeway of genetic information, RNA and proteins hold the same evolutionary secrets as DNA itself. In Pauling's terms, all three are semantides. Crick's dogma summarizes the essence of biological function. Information stored in DNA directs through messenger RNA the formation of proteins, which carry out and regulate the cell's manifold activities.

"Dogma" may seem an odd term to apply to a scientific theory, but Crick chose it consciously. Later in his autobiography, *What Mad Pursuit: A Personal View of Scientific Discovery*, Crick explained that he felt dogma was "more central and powerful" than theory. The religious implications didn't deter him because he "thought all religious beliefs were without foundation."

The elucidation of DNA's structure also provided the explanation for a curious characteristic of DNA discovered years earlier in the 1940s by the Austrian-born biochemist Erwin Chargaff. His principle, which became known as Chargaff's rules, was better understood after Watson and Crick's 1953 discovery.

On learning of the startling success of Oswald Avery, Colin MacLeod, and Maclyn McCarty in identifying DNA as the material capable of altering the genetic makeup of certain bacteria, Chargaff along with many others became convinced that genetic information was encoded in DNA, and in fact that genes were made of DNA. To decipher how DNA could store information, Chargaff studied its composition in a variety of sources, from bacteria *(Escherichia coli)* and yeasts to sea urchins, corn, wheat, grasshoppers, rats, and humans. He found startling differences and one remarkable similarity. The relative amounts of the various nucleobases (A, C, G, and T) that they contained varied enormously from one species to another, but one property was invariant. The number of molecules of G making up a particular DNA is always equal to the number of molecules of C it contains, and the number of molecules of A always equals those of T. These became Chargaff's rules: in all DNA, A always equals T, and G always equals C.

DNA's double-helix structure confirmed the logical basis of Chargaff's rule. Because hydrogen bonds between G and C as well as those between A and T hold its two strands together, the number of As in any molecule of DNA has to equal to the number of Ts and the number of Gs must equal the number of Cs. But DNA's structure makes no demands on what fraction of these hydrogen bonds are G-C or A-T. In fact, these fractions vary enormously among organisms. For example, the percentage of the hydrogen bonds that are G-C bonds in the DNA of the malaria-causing organism *Plasmodium falciparum* is 20 percent. In an antibiotic-producing bacterium, *Streptomyces coelicolor,* it is 72 percent. Even among rather closely related bacteria, the differences are huge, ranging from 25 percent among species of *Micrococcus* to 75 percent in species of *Mycoplasma.* These percentages are uniform throughout a cell's entire genome; they characterize it. From this fact, inheritance can be deduced: finding a stretch of cell's genome with a percentage of G-C bonds that differs from the rest of the molecule is powerful evidence that this stretch of genome is not native to that genome. It must have come (relatively recently in evolutionary time) from somewhere else.

Writing Instructions for Life

By knowing the structure of DNA, researchers could address biology's most fundamental questions, including this: What determines a protein's primary structure? How is the sequence of particular amino acids that compose a particular protein encoded in DNA to make up that organism's genetic code? Clearly, each "word" of the code, later dubbed a codon, must designate a particular amino acid, but how are the words spelled, how long are they, and how are they arranged? Pauling might have stated it this way: In what encrypted form is information passed down the semantide pathway?

This splendid, fundamental, and at the time seemingly impenetrable question was answered just a few years after Watson and Crick's discovery of DNA's structure. The answers were obtained through the unlikely alliance of two very different but complementary modes of probing biological questions: genetics and biochemistry, approaches that lie at the extremes of the spectrum of scientific abstraction.

Genetics, prior to the elucidation of its chemical basis in about 1960, occupied the highest ground of scientific abstraction. Classical genetics drew penetrating conclusions from seemingly disconnected abstract observations. For example, the father of genetics, the nineteenth-century Silesian friar Gregor Mendel, discovered the particulate (gene-based) nature of inheritance simply by observing the lines of descent associated with the colors and shapes of garden peas. Mendel didn't study the genes themselves, nor did he have any idea of what they might be. He observed their consequences, and by abstract reasoning drew conclusions about their causes, i.e., genes. He discovered that inheritance is particulate and unitary, not a continuum of mixing as humans until then had tacitly assumed.

In contrast, biochemical analysis of biological systems lies at the opposite extreme of scientific abstraction. It depends on dealing directly with the object of interest, disassembling it into its recognizable chemical components, and then reassembling only those sufficient to mediate the phenomenon being studied. In the case of cracking the genetic code, both genetics and biochemistry were exploited: a genetic approach revealed the code's syntax, and a biochemical one translated its vocabulary.

The remarkable syntax-revealing experiments are a monument to the power of genetic abstraction. Through them a fundamental law of nature was discovered by the simple act of looking at and counting bacterial virus plaques on Petri dishes. These bacterial virus plaques are small circular clearings on a contiguous lawn of bacteria caused by an epidemic infection, usually started by a single virus particle infecting a single bacterial cell. The experiments are the equivalent of Mendel's in recording the color and shape of flowers and seeds of peas, combined with considerable abstract reasoning.

The illustrious quartet of Francis Crick, Leslie Barnett, Sydney Brenner, and Richard J. Watts-Tobin, working under apparently Spartan conditions in the Cavendish Laboratory of Cambridge University, England, focused on a recognizable mutation, designated rII (*r* for "rapid lysing"), of the bacterial virus, bacteriophage T4. Rapid lysing is an assumed explanation for the larger plaques that characterize the mutation and is used to distinguish its presence or absence. The quartet had no idea of the biochemical basis for the mutation that speeded lysis and increased plaque size, nor did they pursue that question. They knew only that the mutation did alter plaque size and that it changed as well the spectrum of susceptible bacterial strains. These readily observable properties, big plaques versus little plaques, allowed the experimenters to score whether or not a plaque was caused by an rII mutant virus or by its wild-type parent. That's all they needed. They made hypothesis after hypothesis and tested them by examining which treatments caused or corrected the mutation, always as revealed by looking at sizes of plaques on a Petri dish.

First they hypothesized that the language of DNA is read starting from a fixed point on the DNA molecule, that words (codons) are read sequentially in one direction from that point, and that codons are not separated by distinguishing marks, that is, the molecular equivalents of commas or spaces, which were wonderful additions made to western written language by Charlemagne's scholars. If such were the case, they reasoned, removing or adding one base in the sequence would render all subsequent (downstream) codons incorrect. For example, a series of "abc" words would all become "bca" were an upstream "a" removed. And the quartet thought they knew how to remove or add one base at a time to a DNA sequence, namely by treating the phage with a known mutagenic chemical called proflavin. They would test their hypothesis by looking at the size of plaques. Molecular models and the total lack of function of the mutants that proflavin generated suggested such a mode of its action. If that were the case, an addition mutation (+) could be corrected by a nearby removal mutation (−). It would reset the register and cause all subsequent words to read "abc" again. Of course, words between the two mutations would be incorrect, but most of the affected gene would be nearly like its original, an unmutated wild-type ancestor. It would be a "pseudo wild-type," and perhaps it would be functional as long as the two mutations were relatively close to one another. So they re-treated one of their arbitrarily designated addition mutations (+) again with proflavin and were indeed able to recover pseudo wild-type strains as a consequence of their acquiring a presumed minus mutation (−). In this way they generated a set of such arbitrarily designated "minus" mutations. By genetic crosses (effected readily by simultaneously infecting bacteria with two different mutant viruses) they rearranged the minus mutations in various combinations. The combination of two minus mutations produced a totally functionless rII gene, but the combination of three minuses produced a pseudo wild-type strain. That is, removing a complete "bca" by the three minus mutations restored all subsequent downstream words to their correct "abc" form. Thus they discovered the syntax of the genetic code.

The Cavendish quartet, by looking at clear areas in a lawn of bacteria, had solved the essence of the syntax of the genetic code: reading starts from a designated point on a stretch of DNA; it proceeds by reading each subsequent string of three bases as a word (a codon) of the genetic code, and it lacks punctuation.

The logic of simple mathematics supported the researchers' conclusion that the code was written in triplets. If the codons were comprised of only a single nucleobase, that is, if each of the four nucleobases designated a particular amino acid, then codes for only four amino acids could be written. If the codon were two bases long, that is, if each pair of bases could designate a particular amino acid, then sixteen could be designated because there are sixteen possible pairs of four (4×4)—not quite enough to designate twenty amino acids. So mathematics demands that codons must be at least three bases long. There are sixty-four possible ways ($4 \times 4 \times 4$) that the four nucleobases can be arranged in groups of three. Mathematics also argues for a triplet code, and that is what we now call a codon.

A triplet code is indeed much more than adequate. Such a code can contain sufficient information to designate sixty-four possibilities, more than is needed for the twenty amino acids used to make most proteins. Some of the excess encoding capacity (sixty-four minus twenty) of the triplet code serves vital functions, such as to designate where on a cell's long stretch of DNA a particular gene starts and where it ends. Three triplets are used to mark the ends of genes; one designates the beginning. The other excess coding capacity seems to be wasted; mathematics doesn't compromise even with the power of natural selection. In most cases, several nucleobase triplets encode the same amino acid. They are redundant synonyms, and they are the rule. Only two amino acids (methionine and tryptophan) are encoded by a single codon. Ten are encoded by two codons. One is encoded by three codons, and the remaining seven amino acids are encoded by four. Such redundancy offers organisms a choice of which codons it prefers. As we'll see redundancy affects the fraction of G-C bonds in an organism's DNA because the G and C content of redundant codons vary. For example, GUC and GUA are redundant codons for the amino acid valine. If organisms use the GUC codon preferentially, its DNA will be richer in C. In turn, such preferences put a stamp of identity on a cell's DNA. Remarkably, the Cavendish quartet predicted such redundancy simply by looking at plaques.

The Cavendish quartet showed that the codons of the genetic code are three bases long, but what do these genetic words mean? How did a small group of scientists figure out which string of three nucleobases designates which particular amino acid? Knowledge of ribosomes and the role they play in protein synthesis played an important role, along with hard-core biochemistry and diligent effort. Marshall Nirenberg, then in his early thirties, was the groundbreaking leader in these efforts. He is credited with the fundamental experiments in 1961 that led to breaking the code, an achievement that won him a share in the Nobel Prize.

How the code was broken is an interesting story, well worth a diversion. Although reading the code is the essence of a genetic question, its breaking was a triumph of biochemistry quite the opposite philosophically from genetics with its guiding philosophy of reductionism. Take complex systems apart; separate their components; reassemble only those that do the job you're interested in studying. Then you have a simpler system that is easier to manage and understand. No abstraction here.

Nirenberg broke apart intact living cells (he used those of the well-studied bacterium *E. coli* and tested to determine if the resulting debris, a mixture of enzymes, ribosomes, and small molecules, was able to make proteins or even the short beginnings of them, which are called polypeptides. It was not. That disappointment turned out to be an opportunity, critical to his approach to breaking the code. Soon he discovered that the mixture became able to make polypeptides if small amounts of RNA were added. He suspected, correctly as it became clear, that these needed RNA molecules were messenger RNA (mRNA) molecules. At the time, it was hypothesized that mRNA molecules were short-lived carriers of instructions from DNA (genes) to ribosomes, where polypeptides are made. These were the RNA intermediaries mentioned in Crick's central dogma and Pauling's semantide pathway. The message these RNA molecules carried almost certainly was written in the genetic code. If this were indeed the case, Nirenberg realized he could write his own messages by synthesizing specific kinds of molecules of RNA. He could read what they meant by determining which amino acids were incorporated into the polypeptides they designated. That would tell him how the code is written—how the genetic information is encrypted, and what the codons mean.

First he chemically made and added mRNA molecules that were a long string of the nucleobase uracil, a string of UUU codons (polyU) that read the same backwards and forwards, and from any point to any other point in the string. In response to adding this messenger, the mixture of broken *E. coli* cell debris made a polypeptide composed exclusively of the amino acid phenylalanine (polyPhe). In a stroke, this experiment had deciphered one codon of the possible sixty-four: the codon UUU instructs the ribosome and its associated machinery to add phenylalanine to the polypeptide it is making. Similarly, by making and adding polyA as a mRNA, he discovered that AAA encodes the amino acid lysine. Making polyC (CCC) encodes the amino acid proline, and GGG was found to encode the amino acid glycine.

Because Nirenberg was unable to synthesize RNA molecules with repeated codons containing more than a single kind of nucleobase, these four codons were the only ones he was able to decipher using this approach. So Nirenberg, following his biochemical instincts, further reduced the complexity of his system to match his ability to synthesize RNA. Although he couldn't make strings of a particular mixed codon, he could make individual mixed codons. In other words, he could make mixed RNA molecules that were only three nucleobases long. These, of course, were single codons, and so they cannot designate synthesis of a string of amino acids in a polypeptide. They can do something, however. They can start the first step in the process of peptide synthesis by binding specifically to "activated forms" of amino acids (called aminoacyl-tRNAs), the forms that donate amino acids to the polypeptide in the processes of being synthesized on a ribosome. These single codons carried a message that could be read by determining which activated amino acid bound to a specific codon. He found, for example, that the UGU codon bound specifically to the activated form of the amino acid cysteine. Thus he learned that the UGU codon encodes cysteine. Using this approach, he deciphered forty-seven of the possible sixty-four codons.

At this point the torch was passed to Har Gobind Khorana, a native of a small Punjab village then working at the University of Wisconsin. Khorana's approach was more intensely chemical. He had developed the methods by which Nirenberg synthesized the various single codons that he used in his binding experiments. Khorana next synthesized all sixty-four possible codons and used them to do the binding experiments that Nirenberg had pioneered. Some of these experiments gave equivocal results. Then, using a combination of chemical and enzymatic methods, he proceeded to synthesize larger RNA molecules. He chemically synthesized short sequences of DNA, which he used as templates to synthesize enzymatically longer DNA molecules. These longer molecules were transcribed enzymatically into RNA. Under the conditions he employed, such transcriptions were repetitive. Thus, he made repetitive sets of two nucleobases, such as UCUCUC. He made an array of repetitive sets of two, three, and four nucleobases. Then he added these RNA molecules as messengers, as Nirenberg did, and determined which polypeptides were made. The repetitive sets of two generated two alternating codons—for example the repetitive set of UCs generated alternating UCU and CUC codons. In the Nirenberg system, this messenger caused the synthesis of a peptide composed of alternating leucine and serine units. But from this experiment alone it was not possible to know which of the two codons designated which amino acid because he did not know where reading began. Khorana completed numerous experiments with four different two-nucleobase repeats, eight different three-nucleobase repeats, and four different four-nucleotide repeats. From this massive set of data he was able to determine each of the sixty-four possible codons encoded. Sixty-one encoded amino acids, and three encoded "stop," the end of a protein.

The enormous and fundamental challenge of learning to read the genetic code, perhaps Nature's most intimate secret, and determine what takes place on the ribosomal workbench had progressed rapidly. These achievements were recognized in 1968 when Nirenberg and Khorana, along with Robert W. Holley and later the Salk Institute of La Jolla, California were awarded the Nobel Prize in Physiology or Medicine. Holley had determined the structure of the active forms of the amino acids that are added to growing polypeptides, the forms to which single codons bound. This active form of the amino acid is a remarkable molecule; it's the amino acid attached to a molecule of RNA called transfer RNA or tRNA. A better name for transfer

RNA would be "translation RNA" because it translates between the languages of nucleic acids and protein. One portion of the molecule (the anticodon) recognizes and binds to its cognate codon on messenger RNA; another portion accepts the amino acid designated by the codon. Holley was the first to determine the structure of one of these critically important molecules, thereby revealing the molecular basis of the step in the semantide pathway in which information is passed from RNA to protein. Years earlier, Francis Crick had hypothesized the existence of just such a molecule.

The climactic cracking of the genetic code brought on a sort of instant scientific nostalgia, particularly for those who were in the midst of solving the problem themselves. What could be more fundamental than learning how nature writes her instructions for life? Who would not want to be a participant in such an enterprise? Only a very few were able to do so. They quickly spoiled all the fun.

The molecular biologist Gunther Stent, as a tribute to this achievement, famously lamented that there were no remaining equivalent biological challenges worthy of great minds. In his 1969 book *The Coming of the Golden Age,* he posits that progress ends when humanity has learned all it can in a particular field, and with the cracking of the genetic code, that end has been reached in molecular biology. He later conceded that he had vastly overstated the case, but he did make abundantly clear the huge emotional impact on biologists of cracking the genetic code. Certainly it was not an end: as is the case with all major scientific accomplishments, it was a dramatic new beginning. Nature's secrets could now be read in her own language.

Although the genetic code used by *E. coli* had been broken, what code did other organisms use? Soon it became clear that this genetic code was almost certainly universal. The UUU codon, for example, that Nirenberg found to encode the amino acid phenylalanine in *E. coli* in fact encodes the same amino acid in all living things that have been tested, ourselves included. All living things are linked by a common genetic language. The fact of the code's universality is, on its own, a powerful argument for there being only a single tree of life. The code is complex and seemingly completely arbitrary. Life's common ancestor must have stumbled on the code, and all extant creatures inherited it through common descent, the conviction sometimes called the "frozen accident theory." Some biologists, however, make the case for a code that has a certain compelling logic and efficiency that natural selection demands, and therefore insist that the code could have been arrived at independently more than once.

Ribosomes, the workbenches where DNA's encryptions are decoded in the process of making polypeptides and proteins, are also universal. All cells make proteins. All cells contain ribosomes, which are a complex of three RNA molecules and a number of proteins. The protein components vary from one organism to another, but all ribosomes contain three molecules of RNA—a tiny one called the 5S RNA subunit and two other components, a smaller subunit and a larger one. What could be a better candidate for molecules in which the complete evolutionary history of all organisms is written? It seems a lot to ask.

Does Ribosomal RNA Have a Longer Reach?

Is it indeed possible that one molecule, or perhaps any or all of these ribosomal RNA molecules, could survive the vastness of life's diversity and evolutionary antiquity to reveal relationships among all organisms? Could it really help us explain the kinship of whales and bacteria? There was a basis for optimism. Two sets of experiments published almost simultaneously in 1965 suggested that ribosomal RNA (rRNA) does have an extensive taxonomic reach and is heavily endowed with evolutionary inertia, meaning that it changes, but only slowly. Both sets of experiments asked about the similarity of the rRNA molecules of various organisms and if they were more similar than other homologous macromolecules. Do they change more slowly than most proteins and thereby become endowed with a greater evolutionary reach?

Roy Doi, a US citizen born in Sacramento, California who spent three years during World War II in Tule Lake and Heart Mountain internment camps for being ethnically Japanese, along with his colleagues at Syracuse University, took the direct approach of characterizing cells' DNA. He and his colleagues used a method called hybridization, discovered by Julius Marmur, a student in Paul Doty's laboratory at Harvard University in the late 1950s. DNA-DNA hybridization is a disassembling and reassembling in the laboratory that exploits DNA's structure, which is dependent on hydrogen bonds. DNA is double stranded because, as we've noted, its individual single stands are held together by the specific hydrogen bonds that form spontaneously between Gs on one strand and Cs on the other, as well as between As and Ts. These are hydrogen bonds, weak bonds that are easily ruptured by mild treatments such as a modest rise in temperature. When heated, double-stranded DNA separates ("melts") into its two component single strands as the notoriously fragile, linking hydrogen bonds become further weakened. When cooled again, the linking hydrogen bonds on complementary strands of DNA reform spontaneously, recreating double-stranded DNA, a process called annealing. For the hydrogen bonds to reform, the matching G and C as well as the A and T nucleobases have to find one another. Many such encounters are needed to reform a firmly bound, double-stranded DNA, a complicated event that might seem improbable. Annealing double-stranded DNA reforms like a zipper.

Single strands obtained by melting DNA from different but very closely related organisms will also anneal to form new, hybrid double-stranded DNA, but because the two strands in hybrid DNA are not completely complementary, there are gaps in the string of hydrogen bonds. The two strands are held together feebly and can be separated more readily. Such hybrid DNAs have lower melting points (T_m) than their two DNA parents. The decrease in T_m is proportional to the degree of dissimilarity; it becomes a measure of how closely its parental DNAs are related.

If two DNAs cannot form hybrids at all, their parents are more distantly related, but there is no way to measure the distance.. The ability to form hybrid DNA is quite narrowly restricted. In general the DNA encoding most genes can hybridize only if taken from closely related organisms. This restricted reach of hybridization limits its value as a tool for studying the broad ranges of relatedness displayed in the Tree of Life.

Doi and his colleagues were interested in a more sophisticated question. Does the relatedness reach of some genes extend over greater evolutionary distances than for some others? A gene that evolved rapidly would lose its identity over the broad sweep of evolution. They took a slightly different approach to answering this question. Rather than studying DNA-DNA hybridization, they looked at hybridization between a single strand of DNA and one of an RNA molecule (mRNA) copied from it. (Recall the flow of Francis Crick's central dogma, that information passes from DNA through mRNA to proteins, thus a single strand of mRNA could hybridize with a single strand of DNA.) Doi's mentor professor, Sol Spiegelman at the University of Illinois, had shown that this sort of hetero nucleic acid hybridization is possible. It undoubtedly occurs naturally as mRNA molecules are being copied from their encoding genes.

Doi and his colleagues found the reach was variable among various genes. Notably, the reach of ribosomal RNAs (rRNAs) is by far the greatest. Apparently the structure of rRNA evolves quite slowly, a result offering promise that the sequence of its nucleobases might well be able to tell the history of all evolution and the relationships among all organisms.

Julius Marmur, now working at the Albert Einstein College of Medicine of Yeshiva University in New York, came to the same conclusions as Doi. He addressed the question of the reach of rRNA by a completely different route, however, namely by exploiting a process of bacterial genetic exchange called transformation.

Transformation is the simple and direct way that some bacteria exchange genetic material. First, one cell releases some of its DNA into its environment Another cell then takes it up and incorporates that DNA into its genome. By this process of recombination a bit of its own DNA is replaced. Much like the procreative process of fish, no direct contact is made between the two cells. During the genetic exchange, the germplasm floats freely in the environment.

After the free-floating DNA enters a recipient cell, it must be paired with the cell's own resident genome in a process reminiscent of hybridization. This is the first step in the process of incorporation into its new genetic home. The same rules apply. There must be substantial similarity between the two bits of DNA for incorporation to be successful and transformation to occur. So the frequency of transformation of a particular gene is a sensitive measure of the similarity of incoming and resident DNA and the relatedness between the organisms donating and receiving the transforming DNA. A broader range of organisms can be compared using genes that evolve slowly. Marmur and his colleagues came to the same conclusion as Doi: ribosomal RNA changes quite slowly over evolutionary time. It has an exceptionally long evolutionary reach. Its encoding genes can be transformed over a more extended range of relatedness than can other genes. Ribosomal RNA is indeed special.

CHAPTER 4

The Rosetta Stone

Carl R. Woese knew of rRNA's great potential as a candidate Rosetta Stone for evolutionary relationships. He sought methods to mine the evolutionary treasures written in the sequence of its nucleobases. In the early 1970s, when he began his quest, it was not feasible to sequence nucleic acids, either DNA or RNA, by direct chemical assault. He had to find an indirect way to probe rRNA's secrets. His journey took him from the laboratory and work of Frederick Sanger in Cambridge, England to Woese's own laboratory in Urbana, Illinois.

Frederick Sanger received the Nobel Prize for Chemistry in 1958 for determining the structure of insulin. He then turned his attention to finding a way to sequence the nucleobases in DNA. Sanger's method, the dideoxy method, was the one used to sequence the human genome, which some consider to be biology and medicine's greatest twentieth-century achievement. In honor of his method, still used to a limited extent today, Sanger received a second Nobel Prize in Chemistry in 1977. This gave him the unique distinction of being the only person to receive two such awards in chemistry, and perhaps one of the few, if not the only, to refuse a knighthood because he didn't want to be addressed as "Sir."

Sanger's two achievements were connected. Determining the structure of insulin accomplished more than just learning about this vital hormone. It led to an understanding that the string of amino acids composing each protein are arranged in a precise order, in essence the fundamental consequence of the information flow of semantides that can be exploited to determine relatedness. Sanger's studies between the prizes in 1958 and 1977 led to methods of approaching the basic structure of ribosomal RNA and the sequence of the nucleobases it contains.

Using a technique that was straightforward yet innovative, Sanger relied on an enzyme to cut a macromolecule in specific places. He then studied the resulting fragments. Sanger, like Pauling, depended on the precision and specificity by which enzymes cut macromolecules. He chose to study the 5S subunit of ribosomal RNA, the smallest of a ribosome's three component RNAs, and to cut it with an enzyme called ribonuclease T_1, which severs RNA molecules after each G residue that it contains. Then he separated the resulting fragments by electrophoresis, as Pauling did his protein fragments: spotting a bit of mixture on a sheet of cellulose acetate and applying a voltage across it, first in one dimension and then in the other. The electrical attraction for the charged RNA fragments moves each of them over the sheet at its own characteristic rate so that each fragment remains as a distinct spot at a particular location on the sheet. Sanger could visualize the array of spots on photographic film because the 5S ribosomal RNA was obtained from cells that had been cultured in the presence of radioactive phosphate (^{32}P). Radioactive decay exposed the film, revealing where the spots lay. Around twenty spots were detectable as separate entities. Ribonuclease T_1 had sliced the 5S RNA he was studying into about twenty pieces.

Then Sanger took the spots off the cellulose sheets and, by an ingenious series of further enzyme cuttings and electrophoreses, he was able to determine the nucleotide sequence of each of the spots. Because of the precision of ribonuclease T_1, all ended with a G. Some were as long as ten nucleotides (UCUCCCCAUG, for example).

Sanger hadn't sequenced 5S RNA because he had no way of knowing how the fragments were linked in the original 5S RNA molecule, but he had learned a great deal about it, even more than Pauling had learned about the amino acid sequence of hemoglobins by using the same principles. Sanger realized the potential of applying this method to the study of relatedness, ending his 1966 paper with a provocative statement: "This study thus provides a basis for finding the sequence of nucleotides . . . as well as a basis for defining structural differences between related mutants or different species of RNA." Carl Woese immediately realized the same thing.

The technology transfer from Sanger's laboratory to Woese's began when Sol Spiegelman, Doi's mentor and one of Woese's colleagues in the Microbiology Department at the University of Illinois, invited David Bishop, one of Sanger's graduate students, to come to his laboratory as a postdoctoral fellow. Using the methods Bishop had learned in Sanger's laboratory, he was to study the genome of some of the RNA-containing cancer viruses that Spiegelman was investigating.

A year later, in 1969, Spiegelman left Urbana to take a position in New York as Director of the Cancer Institute at the Columbia University College of Physicians and Surgeons. Woese might have been the greater beneficiary of Spiegelman's advancement because he inherited many of Spiegelman's highly skilled team of RNA investigators, adding them to his own team.

Although trained as a biophysicist, not a microbiologist, Woese was fascinated by microbiology's most vexing and intractable problem: the phylogeny, or evolutionary relationships, among bacteria, then probably the darkest black box of phylogeny. He used the essence of Sanger's methods to attack it. Woese's aim was to develop a coherent taxonomy for this group of organisms that had defied less sophisticated approaches.

Woese's team began analyzing 5S ribosomal RNA (about 120 nucleobases long) as Sanger had done, but they quickly realized that this small molecule didn't contain sufficient information for their purposes. So Woese shifted to 16S rRNA (about 1,540 nucleobases long), the middle-sized ribosomal RNA molecule. Like Goldilocks's choice of a chair, it proved to be just right, the ideal compromise between information content and feasibility of analysis.

When he cut the 16S rRNA with the enzyme ribonuclease T_1, as Sanger had done, most nucleotide fragments were quite short, but some contained twenty nucleobases or more. The 16S rRNA from each bacterium Woese examined yielded a different set of fragments, producing a different pattern of spots on the cellulose sheets, undoubtedly reflecting the evolutionary distances among these bacteria. But how could that be determined? As Woese and his group looked at the array of spots from the 16S rRNA from more and more bacteria, their skills at recognizing particular spots (individual RNA fragments) improved.

Recognizable patterns began to develop. Spots containing particular fragments that were six or more nucleobases long occurred only once in the 16S rRNAs from most bacteria. Some became old friends, and Woese recognized some as being special. He came to call them "signatures," or fragments that came only from the RNAs of a similar group of bacteria.

After he had examined the 16S rRNAs from about thirty different bacteria he noticed that there were two kinds of signatures. Some presumably changed slowly during evolution occurred in over 90 percent of the species studied. Others, occurring in only about 65 percent of the bacteria, appeared to have changed more rapidly.

Woese was extremely encouraged by these results. He attributed the difference between rapidly and slowly changing signatures to their location. Those derived from some regions of the 16S rRNA molecule could more readily tolerate change and therefore evolve more rapidly than the others. He compared these types of regions to the minute and hour hands of a phylogenetic clock, offering the hope that the minute hand could reveal rather close relationships and the hour hand more distant ones.

The Rosetta Stone

At about that time, Woese was joined by a postdoctoral fellow, George Fox, with little or no background in biology. Fox, whose background was in chemical engineering, was attracted to Woese's endeavors after having read some of Woese's older papers on a proposed mechanism of how ribosomes might move along a molecule of messenger RNA. In the process of reading the mRNA, the ribosomes would determine the proper sequence of amino acids in the protein being synthesized. The paper sparkled with an innovation and excitement that intrigued Fox.

Soon Fox made two important contributions. He developed what he termed a quantitative similarity index (S_{AB}) to compare differences in the complex array of fragments obtained by cutting a pair (A and B) of 16S rRNAs. S_{AB} is simply the fraction of the total number of 16S rRNA fragments that are shared by a pair of organisms, but the concept of S_{AB} made a vital step toward further progress. It allowed the comparison of two intricate patterns of spots to be reduced to a single number, and numbers can be manipulated mathematically. Fox used the set of similarity indices derived from the data they had studied to construct a phylogenetic tree showing the relationships among these bacteria, the first such bacterial tree that could legitimately make a claim for being phylogenetically based. Moreover, this tree was built with data, with numbers.

At that time I was Chair of the Department of Microbiology at the University of California, Davis I learned of Woese's studies and invited him to conduct a seminar for our department. Having heard of his reluctance to travel, I was pleased when he accepted. The seminar was enthusiastically received, and I was startled and moved to see for the first time in my career a taxonomic tree for bacteria based on actual phylogenetic data with a quantitative basis.

I was also relieved. I had become the lead author of *The Microbial World,* a textbook initiated by Roger Stanier in 1957 that had played a pivotal, although sometimes controversial, role in forming opinions about bacterial identity and taxonomy. Stanier and his two departmental colleagues at Berkeley, Michael Doudoroff and Edward Adelberg, had written the first three editions. When Michael Doudoroff passed away in 1975, Stanier asked me to fill in for the fourth edition. By the time the fifth edition was due, Stanier was too ill to participate, and Adelberg no longer wanted to be involved. So I assumed the responsibility for writing it, asking Mark Wheelis and Page Painter to give me a hand. Although Stanier was unable to contribute to its rewriting, many of his words and even more of his style were carried over from previous editions. I felt the weight of having inherited *The Microbial World* and its dominant position as an arbiter of rational bacterial taxonomy. Woese's approach and results were enormously comforting. Perhaps in future editions we could write about a rational, phylogenetic taxonomy for bacteria.

At a national meeting of the American Society for Microbiology, Ralph S. Wolfe told me with considerable excitement that he and Carl Woese had made a major discovery. Methane-producing bacteria, methanogens, which he had been studying for years, were quite unlike any other bacteria. I thought, but didn't say, that I already knew that. Methanogens are special in so many ways, as Wolfe and Robert E. Hungate, a colleague in my department at the University of California, Davis and a pioneer in studying methanogens, had already shown. I didn't appreciate the full impact of what Wolfe was telling me: methanogens, as revealed by Woese's methods, were as phylogenetically distinct from other bacteria as they were from animals. When I learned of the trail of events leading to Ralph's enthusiastic comments, however, the impact of the finding became apparent.

Making spectacular progress toward bringing a rational approach to bacterial taxonomy by analyzing 16S rRNA from readily available species of bacteria, Woese had sought sources of more varied bacteria to study. He consulted his friend and colleague Ralph Wolfe. Despite having a lukewarm attitude about "bug sorting" as a serious scientific enterprise, feeling it was based on hunches rather than hard data, Wolfe suggested that Woese apply his methods to investigating the bacterial genus *Bacillus*. I suspect that Wolfe considered this undertaking to be a test or evaluation of the usefulness of Woese's approach. The genus, *Bacillus,* is one of the few well-defined groups of bacteria on morphological and metabolic grounds. Its members are crisply identifiable: Gram-positive, needing access to oxygen (though some can substitute nitrate for oxygen), and able to form extremely hearty, highly resistant, and biologically unique cells called endospores. Woese's results from his studies on the 16S rRNA of members of the genus *Bacillus* converted Wolfe to believing that "bug sorting" could indeed have a sound scientific basis. It could, Wolfe now believed, make important contributions to microbiology, perhaps to all biology.

Wolfe then proposed a greater challenge: Woese should investigate those odd microbes called methanogens that were the focus of Wolfe's own biochemical research. These microbes, which everyone considered to be bacteria, were well known for their bizarre properties. They obtain energy for growth and multiplication by making one gas, methane (CH_4). They accomplish this by reducing another gas, carbon dioxide (CO_2), by yet a third gas, hydrogen (H_2), or in some cases by certain other reducing agents. Wolfe and his colleagues had also discovered that methanogens are biologically unique in other respects. They contain a small molecule, Coenzyme M, which is a member of a particular class of the compounds called coenzymes. Coenzymes are required partners for the activity of certain enzymes. Coenzyme M partners in the central core of reactions that form methane. It is essential and responsible for the approximately billion metric tons of methane that methanogens produce annually in many oxygen-free environments, including the intestines of some of us.

Coenzyme M occurs only in methanogens and nowhere else in nature. Moreover, the cell walls of methanogens lack a structural macromolecule called peptidoglycan that is found in the overwhelming majority of bacteria. Methanogens, at least in these respects, are unlike other known organisms, and for this reason alone worthy of careful investigation. But methanogens are exasperatingly difficult to study.

Although they abound in many oxygen-free, nutrient-rich natural environments, they were extremely difficult to culture in the laboratory, at least with the equipment available at the time of Wolfe and Woese's studies. Methanogens are so rigorously oxygen phobic that they are unable to survive even a brief exposure to oxygen. Culturing methanogens was an enormous technical challenge that only a handful of microbiologists had then been able to master.

Robert Hungate was one of the masters, the leading one. The first student of C. B. van Niel, the mid-twentieth-century microbiologist responsible for key discoveries about photosynthesis, Hungate had developed a set of methods that came to be known as the "Hungate Technique"; it permitted the successful culturing in an ordinary microbiological laboratory of methanogens from the rumen of cows and other anaerobic environments suited to their intensely oxygen-averse lifestyle. There are a surprising number and variety of methanogen-suitable environments in our oxygen-permeated planet. One such location is mud at the bottoms of quiet ponds, where oxygen-utilizing microbes scrupulously deplete the gas to minuscule concentrations. There, methanogens flourish and are responsible for the bubbles (of methane) we often see rising from such ponds. Another location is our own intestines. In the rumen of cattle and other related animals, methanogens produce the considerable quantities of methane that cattle release by belching—a source of methane with considerable environmental impact. Belching cattle collectively are thought to produce about a quarter of the world's methane, which is a twentyfold more powerful greenhouse gas than CO_2. The methane cattle belch contributes significantly to global warming. Most of us humans also carry methanogens in our gut and are responsible for producing methane as an intestinal gas, but presumably we're not in the same league as cattle. I'm not aware of any study assessing the contribution of human intestinal gas to world methane production and hence global warming. Considering our prodigious numbers, however, it must be significant.

The principle of the Hungate technique for culturing methanogens is straightforward: carry out all manipulations under a protective stream of oxygen-displacing nitrogen gas. Success depends on meticulous attention to the details. All traces of oxygen must be excluded from the protective layer of nitrogen, a fact that Hungate learned early and painfully in his studies. When Hungate began his investigations, he was unsuccessful in his attempts to propagate methanogens in laboratory culture. He soon learned by trial and error that commercially available bottled nitrogen gas contained minute amounts of oxygen—enough to kill the exquisitely sensitive methanogens he set out to culture. To use bottled nitrogen he had to render it scrupulously oxygen free, which he accomplished by passing the bottled gas through a long glass tube packed with metallic copper heated by a string of Bunsen burners, an odd Rube Goldberg–style contraption that was housed in the laboratory down the hall from mine. The heated copper reacts with and effectively removes any residual oxygen from the stream of nitrogen.

Wolfe had learned the Hungate technique from Marvin Bryant, a former student of Hungate who had moved to the dairy science department at the University of Illinois at Urbana-Champaign. With the help of Meyer Wollin and Eileen Wollin, also in the dairy science department, Wolfe had become able to cultivate methanogens in mass culture, enough to supply the large quantities of cells that he needed for his studies on the biochemistry of methanogenesis.

Discovering the Third Domain of Life

Wolfe's mastery of the Hungate technique meant that he could supply cells of the methanogen *Methanobacterium thermoautothrophicum*, the species he was using in his biochemical studies, to Woese in quantities sufficient for him to examine its 16S rRNA, an offer which turned out to be a paradigm changer for all of biology.

The pattern of RNA fragments from methanogens that Woese saw was so startlingly different from those of any of the bacteria he had been studying that he thought something must have gone wrong with his procedure. Missing were certain signature fragments that he had always seen when he examined 16S rRNA from bacteria, and there were new signature fragments that he had never seen before. He asked Wolfe to supply him with a new batch of cells. The new batch showed the same extremely unusual pattern. This time Woese realized that it was the source of the cells that was so extremely odd. He presciently told Wolfe, "These cells aren't even from bacteria." Wolfe assured him that they were. They looked like bacteria. Their cells had prokaryotic structure. Then Woese confidently proclaimed, "They're not related to anything I've seen." In retrospect, Wolfe realized that at that moment the third major branch of the Tree of Life, later called the archaea, had been discovered. Methanogens were not bacteria. They were something completely different. Woese trusted his results and immediately sensed the importance of his discovery. According to Fox, "When he realized this wasn't a mistake, he just went nuts. He ran into my lab and told me we had just discovered a new form of life."

Wolfe, Woese, and Fox in collaboration with Linda J. Magrum and William E. Balch then cultured and examined the small subunit

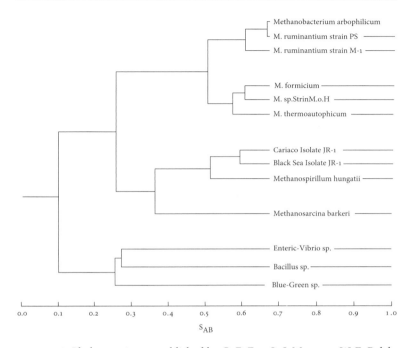

FIGURE 8. Phylogenetic tree published by G. E. Fox, L. J. Magrum, W. E. Balch, R. S. Wolfe, and C. R. Woese showing the differences between methanogenic bacteria and other well-known bacteria. These results led to the discovery of Archaea as a distinct domain of life.

rRNA of ten species or isolates of methanogens then available to them. (Collectively, 16S and 18S rRNA are called small subunit RNA.) The results were spectacular. Using a similarity index (S_{AB}) developed by Fox as a quantitative estimate of relatedness between these various methanogens and bacteria, they constructed a small phylogenetic tree of methanogen relatedness.

The methanogens formed a distinct branch of this tree, a tight set of clusters of relatedness that all had shared similarity indices of 0.25 or greater. Three bacteria they had previously studied formed another branch of the tree, also with shared similarity indices of 0.25 or greater. But the similarity indices between the two branches, between the other bacteria and the methanogens, was slightly less than about 0.1. By this method of reckoning, the methanogens were indeed quite different from the bacteria Woese had studied and possibly from all other organisms as well.

In 1977 a paper with an unassumingly modest title, "Classification of Methanogenic Bacteria by 16S Ribosomal RNA Characterization," summarized these results and was published by the journal *Proceedings of the National Academy of Sciences USA*.

Woese believed this was a critically important paper, and he didn't want it to go unnoticed. He was somewhat sensitive along these lines. It was his nature, perhaps augmented by his resentment of what he perceived to be the scientific neglect of a previous publication. In this earlier paper, he had proposed a mechanism by which a ribosome might move down the message in the process of protein synthesis by repeated ratchet-like shifts of the ribosome's molecular structure. His approach was innovative, complex, imaginative, and later proven incorrect, although as Woese suggested the mechanism conceivably could have functioned in the evolutionary past.

Woese did not take such scientific oversight lightly. I recall his comments at a small dinner in Berkeley following the 1993 Stanier Memorial Lecture. Woese's lecture was generally well received in spite of his having been bitingly critical Stanier himself. Why, Woese asked over dinner, had the scientific community ignored his early innovative idea? Even twenty-three years after the original proposal, with abundant scientific acclaim for his achievements since, Woese could not forgive. As the topic lingered in the conversation, from across the table the microbiologist Gunther Stent ended it abruptly by pointing out that sometimes the most dazzling new ideas turn out to be crazy.

The Rosetta Stone

So for whatever reason, immediately prior to publication of the paper on the methanogens, Woese issued a joint press release with the National Aeronautics and Space Administration and the National Science Foundation, both of which had supported Woese's research. The press release described the essence of the paper's findings: the concept that the archaea constitute the third major branch of the Tree of Life. Wolfe, a coauthor of the paper, later speculated that the press conference in fact delayed general scientific acceptance of the significance of their discovery by a decade. This was probably an overstatement, but misinterpretation by the media certainly had considerable negative impact.

Phone calls from reporters seeking a fuller explanation of the significance of the press release soon followed. Woese then admitted to Wolfe that he had told the reporters that the two of them had discovered a third form of life. This dramatic statement, enigmatic but tantalizing to the reporters, brought the press release to the front pages of newspapers on November 2, 1977.

The local newspaper in Urbana, Illinois, the *News-Gazette,* reacted somewhat mildly with a banner headline, "Ui Research Breakthrough Identifies Third Life Form." The *New York Times* also ran a restrained front-page headline, "Scientists Discuss Form of Life that Predates Higher Organisms," alongside a photo of Carl Woese with his Adidas-clad feet atop his desk. The *Chicago Tribune* had probably the most sensational presentation with a banner headline, "Maybe Oldest Life Form Here. Martian Like Bugs On Earth?" The accompanying article explained incoherently, "No fossil traces of the bacteria *Methanobacterium thermoautrotrophica* have been found because at the time, no rocks had yet formed."

Naturally the scientific community reacted negatively, perhaps with a touch of jealousy, to this prepublication publicity and to such grossly distorted, sensational representation of science, which read like second-rate science fiction. Prestigious scientific colleagues, including the Nobel Prize winner Salvador Luria, a former colleague at the University of Illinois, phoned Wolfe, warning him for the good of his scientific standing to disassociate himself immediately from these crackpot claims. Wolfe pleaded with Luria to read the journal article and not the newspaper reports.

In dramatic contrast to the press coverage, the journal publication drew only a moderate, thoughtful conclusion: "It would appear that methanogens ultimately may have to be classified as a systemic group distinct from other bacteria (inclusive of the blue-green algae)," along with the modest speculation that "Methanogens might have existed at a time when an anaerobic atmosphere, rich in carbon dioxide and hydrogen, enveloped the planet and, if so, could have played a pivotal role in this planet's physical evolution."

One month later, Woese and his colleague George E. Fox took a bolder step, far beyond their initial aims of bringing phylogenic order to bacterial taxonomy and the methanogens. In a paper titled "Phylogenetic Structure of the Prokaryotic Domain: The Primary Kingdoms," also published in the *Proceedings of the National Academy of Sciences USA,* they extended their comparison of the 16S rRNA of methanogens to the equivalent molecule (18S rRNA) of eukaryotes. Thus, they compared these molecules from representatives of all major groups of organisms and found that by this means of judgment, methanogens were quite different from all other living things, so much so that they must constitute a distinct biological group. On the basis of these results, Woese and Fox suggested that the biological world consisted of three major groups, which they proposed should be called primary kingdoms or urkingdoms. These, they felt, ought to be termed the eubacteria (including all bacteria except the methanogen group), the archaebacteria (the methanogens), and the eukaryotes (all other organisms from microbes such as yeasts and protists to humans).

The Rosetta Stone

Woese and Fox came to these monumental, paradigm-altering conclusions on the basis of what seemed to many scientists to be extremely meager data. Using their system of similarity indices of small ribosomal subunits, they had compared three eukaryotes (a yeast, duck weed, and a mouse sarcoma cell line), six bacteria, and four methanogens. The similarity indices between pairs of organisms within each of the three groups (eukaryotes, bacteria, and methanogens) all exceeded 0.20; none of the indices between pairs of organisms from different groups was greater than 0.13.

In this paper Woese and Fox had proposed that we change completely our view of the structure of the biological world on the basis of tiny differences among a small set of numbers. These differences were represented by no more than the arrays of spots on cellulose sheets. That didn't seem to satisfy Carl Sagan's famous dictum, "Extraordinary claims require extraordinary evidence." But Woese and Fox's proposal, largely intact, has stood the test of time and floods of suspicious scrutiny. Clearly, their contribution had Nobel Prize–worthy impact, but curiously no award came. Could the notorious press release have been a factor? Carl Woese, age eighty-four, died of pancreatic cancer in 2012.

Woese and Fox did bravely satisfy another important scientific dictum, one advocated by, among others, a technical director named Max Gable for whom I used to work at the DuPont Company. This dictum? "Believe your data." Max readily cited a list of important discoveries that were overlooked or delayed when unexpected data were discarded because they presented too blatant a conflict with conventional wisdom. Woese and Fox did believe their data, meager as it was, challenged conventional scientific wisdom, and fundamentally changed our view of the living world.

The specter of the taxonomic principle of balance also confronted Woese and Fox's proposal. Throughout most of taxonomic history, two kingdoms—Plantae and Animalia—had been established to accommodate groups of quite different organisms, large groups of them. The methanogens, which Woese and Fox said deserved their own urkingdom, were according to their data indeed quite different, but at the time only a dozen or so representatives were known. The other two urkingdoms they proposed (bacteria and eukaryotes) each contained millions of probable (in the case of bacteria) and well-established (in the case of eukaryotes) representative species. Was there any balance or indeed logic in a taxonomic scheme that squared dozens against millions? Did the methanogens more properly constitute a small collection of prokaryotic oddities rather than a new urkingdom?

Woese's three-kingdom system struck some highly respected taxonomists, including Ernst Mayr, the father of the concept of the biological definition of species, as unbalanced and imperious, a way to magnify the importance of Woese and Fox's results. In a paper published in the *Proceedings of the National Academy of Sciences USA* in 1998, Mayr was initially effusive in his praise of Woese's accomplishments: "Carl Woese's discovery of the archaebacteria was like the discovery of a new continent," but then he added his cautious doubts. "Where should one place this new group of microorganisms?" Mayr felt that the shared cell structure (prokaryotic) between bacteria and archaebacteria, distinguishing them from all other organisms, was so powerful that they should remain linked taxonomically, that there should be only two recognized kingdoms in the biological world, Prokaryota and Eukaryota. Mayr supported his opinion with a scholarly discussion of the principles of taxonomy. The essence of his opinion was disbelief that archaebacteria and eukaryotes should be assigned equal rank. They just didn't seem to him to have equal standing. As he said, "One of the basic principles of good classification is the principle of balance, which states that the retrieval of information is greatly facilitated if the taxa at a given categorical rank are, as far as

The Rosetta Stone

possible, of equal size." Mayr also objected incisively, and somewhat caustically, to Woese's proposed name change of archaebacteria to archaea. Summarizing evidence that archaebacteria might not be Earth's oldest inhabitants, he charged, "Woese renamed them ... retaining the inappropriate component—archaea—and discarding the informative component—bacteria—which revealed their prokaryote nature."

Woese responded rather emotionally one month later in the same journal. Failing to grant equal status to archaea, bacteria, and eukaryotes was not just an issue of taxonomy, it revealed a fundamentally incorrect attitude about biology itself. As he put it, "To return to the prokaryote-eukaryote dogma (with its lingering false connotations) would have a similarly negative effect now, once again on microbiology, but this time too on the study of evolution—both of these fields currently in states of revolutionary development." To Woese, apparently, the very future of the science of biology depended on the taxonomic status accorded the archaea. One would think this a remarkably hard case to make.

Woese and Fox did, however, concede the startling numbers imbalance, but they bravely predicted, on the basis of no evidence that I know of, that more members of their newly proposed urkingdom of archaebacteria would be found. They mentioned extreme halophiles (microbes, then considered to be bacteria, that thrive in extremely salty environments, including saturated brine and rotting salted fish) as possible candidate members of the archaea because halophiles, like the methanogens, lack the macromolecule peptidoglycan, which forms the cell walls of most bacteria.

Time proved Woese and Fox to be right on both scores. The extreme halophiles did indeed prove to qualify as members of the new urkingdom, and imbalance of numbers has been dramatically reduced. Subsequently, many other new members of the archaea have been found. To this date, the archaea urkingdom continues to grow as new representatives are discovered. Now the archaea are commonly subdivided into phyla whose members collectively were recently estimated to constitute about 20 percent of the world's biomass. Certainly the proposed urkingdom has now refuted the charge of imbalance.

As Woese's archaebacteria became more thoroughly studied throughout the 1980s and 1990s, and the profound biochemical, metabolic, and ecological differences between "eubacteria" and "archaebacteria" became increasingly apparent, using the root term *bacteria* in both of their names, in spite of Mayr's later criticisms, began to seem inappropriate. It emphasized what Woese considered to be a misleading similarity, suggesting that these two very different groups of organisms were merely two different forms of the same sorts of creatures.

In 1990 Woese, in collaboration with Otto Kandler and Mark Wheelis, revisited the issue of names. Kandler had played a leading role in expanding the world's knowledge of archaebacteria, and Wheelis, a noted microbiologist, had been a coauthor of *The Microbial World*. The three proposed that the "bacteria" component of the word "archaebacteria" (which briefly, as a nod to Latin linguistic propriety, had become "archaeobacteria") be dropped in favor simply of "archaea" (from the Latin meaning "ancient"), thus also rendering the *eu-* of "eubacteria" superfluous. Their proposed three major divisions of the living world were Archaea, Bacteria, and Eukarya (eukaryotes). Woese was passionate about names as well as many other things, as I learned in 1987 when I was the editor of *Microbiological Reviews*. He had submitted a review, "Bacterial Evolution," which later proved to be a classic. He told me that if house style required him to spell archaebacteria with an *o*, he would withdraw the paper. I assured him he could spell it as he chose.

Woese, Kandler, and Wheelis also proposed a new term for these three major divisions of life: domain. The term *kingdom*, which biology had used for so long, could be preserved as a subdivision of domain. Thus the domain Eukarya would be subdivided into four kingdoms: plants, animals, fungi, and protists. Surprisingly these suggestions have been accepted without too much argument by most biologists.

Thus, Woese, Kandler, and Wheelis christened the trunks and the major branches of the Tree of Life.

CHAPTER 5

From the Tree's Roots to Its Branches

Identifying a cluster of branches, of course, does not describe a tree. Nor does the identification of Bacteria, Archaea, and Eukarya as being distinct but related groups describe a complete Tree of Life. A tree has roots leading to a trunk where the branches join. Locating the root of the Tree of Life, the point where it all began, posed a seeming impenetrable conundrum that lingered into the late 1980s. The point of origin of a group, where it branches away on its own, can only be identified by comparing that group to a related group. Otherwise it just floats independently. For example, the origin of animals can be identified as when they branched away from the line leading to plants. In other words, determining origins of groups depends on comparisons to outside groups. That's the dilemma for determining the root of the Tree of Life. There is no outside group. Life is all there is.

In 1978 Robert M. Schwartz and Margaret O. Dayhoff offered an innovative solution to this quandary, but not until 1989 had sufficient information been accumulated for it to be applied. Schwartz and Dayhoff suggested that each of a pair of duplicated genes acts as an outside group to the other, and prokaryotic genes readily duplicate. As DNA is replicated, mistakes leading to duplications of stretches of DNA or of genes occur at an extremely high frequency. At any one time each gene in a population of bacteria has a duplicate copy at a frequency of about 10^{-5}, that is, one in a hundred thousand cells carries two copies of each gene. But duplicated genes are also eliminated at a high rate by recombination. Duplications are maintained long term only when having two copies of a gene offers a selective advantage to the cell that carries them. Usually that happens when the two copies evolve to fulfill different but essential or advantageous functions. A pair of genes evolving by this route are said to be paralogous. A tree based on descendants from a pair of paralogous genes would branch where they were formed. That's the reasoning that two sets of investigators used to find the root of the Tree of Life. Of course they had to find a pair of paralogous genes that formed near life's beginning, that is, on the trunk near the root of the Tree of Life. In 1989 two groups of investigators identified such pairs of genes and published their results almost simultaneously in the *Proceedings of the National Academy of Sciences,* one in September and the other in December.

The September group consisted of thirteen authors from four institutions. The lead author was Johan Peter Gogarten. Their goal was to study two special forms of enzymes called ATPase. Cells use most forms of ATPase to decompose ATP (adenosine triphosphate) into ADP (adenosine diphosphate) and phosphate. ATP is the currency of metabolic energy used to drive many metabolic reactions that could not occur in its absence. The special forms that Gogarten and his colleagues studied are located in membranes and catalyze the reverse reaction, the synthesis of ATP from ADP and phosphate ion. A flow of hydrogen atoms (H^+) through the membrane drives this reaction. The F form of these special ATPases is located in the plasma membrane, and the V form in the vacuolar membrane.. A comparison of the two forms revealed a similar sequence of their constituent amino acids and their functions. Gogarten and colleagues then realized that these enzymes are paralogous. And because at least one of each form of these ATPases is found in each of the three branches of the Tree of Life, their duplication must have occurred in the trunk of the tree before branching occurred. Constructing a tree based on these paralogous enzymes produces a tree (see Figure 3) with two initial branches. One branch is Bacteria and the other Archaea and

Eukarya, which again splits, each to form its own branch. By this interpretation, the archaea and eukaryotes are sister groups.

The December group, Naoyuki Iwabe and colleagues, studied a different pair of paralogous enzymes, elongation factor Tu and elongation factor G, both of which play similar but distinct roles in protein synthesis. These enzymes, like the ATPases, are also found in all three branches of the Tree of Life. A Tree of Life based on them mimics the one found by the September group, with two branches, Bacteria and a combined trunk of Archaea and Eukarya. This latter branch subsequently splits, so that Archaea and Eukarya are sister groups.

This "rooting" of the Tree of Life has been broadly accepted, although the recent discovery of a new group of archaea, the Lokiarchaeota, brings this deep branching of the Tree of Life into question.

A Taxonomy of Bacteria and Archaea

In spite of his discovery of a third domain of life, Woese returned to his original goal of developing a rational, phylogenic taxonomy for bacteria, but now with the addition of archaea. He continued to rely on differences in small subunit ribosomal RNA as an index of relatedness. As methods of conducting this research improved dramatically, more and more microbiologists began to participate. The method and the results became near-dogma for microbiology. Prokaryotic classification, or "bug sorting" as Wolfe and others had called it, had become a quantitative science. Woese's methods became the fundamental tools of microbial classification and phylogeny. Its arrival at full microbiological legitimacy was signaled by its incorporation into *Bergey's Manual*.

Microbiologists worldwide look to *Bergey's Manual of Systematic Bacteriology* by David Hendricks Bergey as a guide to the accepted scheme of bacterial classification. The second edition of the manual, published in 2001, was based on the results of small subunit RNA comparisons. For the first time a scheme for classifying prokaryotes looked like schemes that have been used to classify plants and animals. A legitimate semantide-based scheme of taxonomy had been developed for the prokaryotes. The group was divided into two domains: the Archaea and the Bacteria. Each of the domains was subdivided progressively into phyla, classes, orders, families, genera, and species. The great goal of bacterial taxonomists, a rational, scientific, phylogenic classification scheme for bacteria and archaea, had once been thought unattainable but now had been reached.. Of course, the details of the scheme will continue to change, but its basic structure has been established.

Molecular methods have become the dominant way to analyze the composition of large mixed populations of bacteria and archaea, such as those that live in and on us. In these studies the results are usually presented in terms of phyla rather than species. There are at present about fifty recognized phyla of bacteria and around twenty of archaea, many of which contain no species that have yet been cultured.

Naming the Major Divisions of Life

Dividing up the biological world into three domains proved surprisingly controversial. Throughout biology's taxonomic history, there's been a curious emphasis on dichotomy in assigning names to groups of organisms. Naming within families of plants or even phyla of animals has been relatively uncontroversial. Specialists of these groups are presumed to know their business, and other biologists usually don't object. Also there has been a certain level of stability at the middle and lower levels of taxonomy. Names are changed relatively infrequently.

But arranging these relatively smaller groups into all-inclusive, larger ones has had quite a different history, particularly with regard to organizing groups of microbes. This activity has been a long-standing intellectual contact sport among biologists. A glimpse at the history of microbe naming reveals biology's chronic frustration when dealing with microbes, and a closer look brings into focus important questions about the pathways through which these major groups evolved from a single common ancestor. Up until Woese, Kandler, and Wheelis published their results, none of the major organizational schemes were based on data. Possibly for that reason, these schemes are particularly revealing records of the progression of biological thought.

As we've discussed, the most widely accepted scheme was the assignment of all living things to one or the other of two biological kingdoms: plants and animals. That's what Linnaeus did in the early eighteenth century. (He added a third kingdom for the nonbiological world of minerals.) As the development of microscopy led to the discovery of increased diversity among microbes, individual groups of them were assigned to either kingdom largely on the basis of movement (these went with the animals) or possession of the green pigments of photosynthesis (these went with the plants). Some taxonomically awkward microbes are green and mobile. They were split. Most were assigned to the animal kingdom but some went to the plants.

In the nineteenth century attempts were made to deal forthrightly with this awkwardness. It became increasingly clear that known microbes are neither plants nor animals. They are different and they ought to be designated as such.

In 1886 Ernst Haeckel, the handsome, but controversial German biologist, offered a solution. Passionate and artistically gifted, Haeckel was also a bit of a linguist. He coined the terms *ecology, phylum, phylogeny,* and even *stem cell.* He supported Darwin's controversial views, and he is also known for the much-cited, catchy phrase that created an analogy between the pattern of embryological development and evolutionary flow, namely, "Ontology recapitulates phylogeny." Faced with the difficulty of classifying motile green microbes, he recommended that single-celled creatures should have a kingdom of their own, and he baptized it "Protista," another Haeckel neologism that is still a part of modern biological vocabulary. *Protos* is Greek for "first." "Protista" is a superlative form, implying "the very first." Such naming offered clear evolutionary implications as well as increased taxonomic logic.

Haeckel went beyond giving microbes their own name. He indicated how he thought they were related to macrobes by sketching a three-kingdom Tree of Life made up of Plantae, Animalia, and Protista. Interestingly, he positioned Protista as the central branch of the tree). In spite of the implications of the term, according to Haeckel's tree this group (microbes) did not give rise to either plants or animals. Haeckel apparently believed they constituted an independent evolutionary line sprouting from the tree's roots. Haeckel's tree clearly expresses his great admiration for Darwin. Darwin acknowledged Haeckel's contribution to the success of his work with the comment, "you are one of the few who clearly understands Natural Selection." Robert J. Richards, historian of science and medicine, suggests that "prior to World War I more people learned of evolutionary theory from the voluminous writings of Charles Darwin's foremost champion in Germany, Ernst Haeckel, than any other source including the writings of Darwin himself." Indeed, Haeckel's writings about Darwinism outsold Darwin's own writings tenfold and more.

Haeckel also recognized the uniqueness of bacteria (archaea, of course, were then unknown). He could see that they are as fundamentally different from other microbes as they are from macrobes. So he separated them into their own distinct subgroup and unsurprisingly invented a name for them, "Monera." "Monera" was yet another

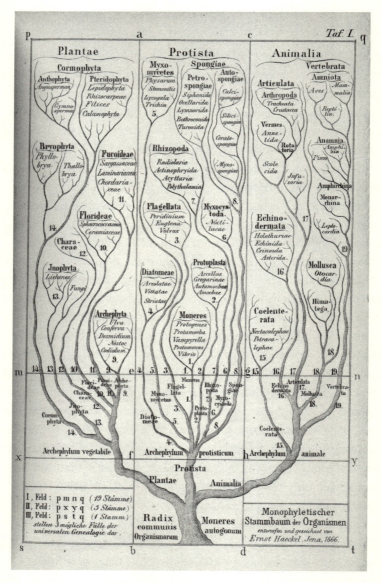

FIGURE 9. Ernst Haeckel's Tree of Life, in which he divides living things into three kingdoms, including a new one, the Protista, which includes all microbes. Bacteria were assigned to a subgroup of the Protista, which he dubbed Moneres.

term invented by Haeckel that he based on the Greek word *moneres*, meaning "single" or "solitary." In this group were protists that lacked a defined nucleus, the group that contains bacteria plus the archaea, the prokaryotes. This tripartite scheme satisfied most biologists for about fifty years.

Then in 1938 Herbert F. Copeland, a little-known biologist at Sacramento Junior College, wrote an imposing, scholarly, thirty-seven-page review of biological taxonomy in the *Quarterly Review of Biology*. Lack of defined nucleus, he noted, was such a profound distinction that such organisms deserved a kingdom of their own. Thus in 1938 he proposed that we ought to recognize four biological kingdoms: plants (Plantae), animals (Animalia), the protists (Protista), and the Monera. His scheme was widely accepted and incorporated into biology textbooks. Kingdom inflation continued, although Copeland kept the fungi within Protista.

The fungi have long been a problem. Haeckel had equivocated about their rightful status. Nonphotosynthetic, they are unlike plants. Certainly, they don't appear to resemble animals in any respect, and their multinucleate, tubelike cell structure starkly distinguishes them from the unicellular protists. Haeckel briefly assigned the fungi to the protists before returning them to the plant kingdom.

Then in 1969 Robert Whittaker, a distinguished plant ecologist, offered what seemed to many to be a logical solution to the issue of the fungi. He gave them their own kingdom, and there were now five biological kingdoms: Monera, Protista, Plantae, Animalia, and Fungi. This five-kingdom classification caught on. Until the revolution of macromolecular sequencing, Whittaker's proposal was the most widely accepted scheme for classifying living things. The system benefited from the enthusiastic support of Lynn Margulis, who famously advocated the theory that mitochondria are captured bacteria, living inside eukaryotic cells and performing aerobic respiration for them. For the rest of her life, Margulis advocated passionately and convincingly to many in favor of Whittaker's five-kingdom scheme.

Prokaryotes and Eukaryotes

About a decade or two before Whittaker proposed his five-kingdom scheme, the electron microscope proved to be an important tool for classifying organisms. In spite of the vastly increased power of magnification the electron microscope offered, from the few thousand-fold of the light microscopes to a few millionfold, the first electron micrographs of bacteria were disappointingly uninformative, adding very little to our understanding of their cellular structure. Early images of rod-shaped bacteria, for example, resembled breakfast sausages, and cocci resembled marbles. Both of these looked about the same under an ordinary light microscope. Because an electron beam cannot penetrate objects even as small as a bacterial cell, the electron microscope revealed only the surface structures of intact bacterial cells, despite the tremendous increase in magnification. These early EM images provided almost no information about intracellular bacterial anatomy because they couldn't see inside the cell.

In 1949 Richard Baker and Daniel Pease achieved the remarkable: they cut a bacterial cell into 0.1 micrometer-thick slices. These ten thin sections could now be penetrated by an electron beam. The images they obtained were stunning. The bacterium's interior looked nothing like the insides of a plant or animal cell., Plant and animal cells, large and not terribly dense, have long been examined in detail using ordinary light microscopes. The images of the bacterium, *Bacillus megatherium* (now spelled *megaterium*), that Baker and Pease obtained were remarkable for what wasn't there. Absent were any of the elaborately differentiated internal structures typical of the cells of plants and animals. Previous observations with light microscopes on stained bacterial cells had established that they lacked a defined, membrane-bounded nucleus, but the monotonous uniformity of the bacterial cell's drab grey interior, revealed in these thin sections, was new and startling.

Soon many electron micrographs of varying quality became available. Most added to our understanding of the intracellular structure of bacteria, but some were misleading. Preparing specimens to be sliced and desiccated so they could be placed in the vacuum that the electron beam requires often distorted and changed the specimens, obscuring their features or even adding artifactual detail. It was not always clear at first that it was the action of preparing the cell that created the artifact. For example, apparent invaginations into cells were often seen in the membranes of certain Gram-positive bacteria, usually those in the genus *Bacillus*. They were given the name "mesosome" (intermediate body) and their function was a matter of speculation. Quite plausible, even convincing explanations for their cellular role were offered. Textbooks pictured and discussed them. But mesosomes turned out to be artifacts caused by certain of the chemicals used to prepare samples for viewing with the electron microscope. When new and more sophisticated methods of preparation were developed, mesosomes disappeared from micrographs and microbiology.

In spite of the hazards of distortion during preparation, many quality thin-section electron micrographs of bacteria became available. Of those who produced quality micrographs early in their history, two quite different scientists stand out in my memory: a German-born researcher with medical training, Carl Robinow, who left Germany just before World War II and moved to Canada after the war, and Eduard Kellenberger, who trained as an atomic physicist at the Swiss Federal Institute of Technology in Zurich. Robinow was a cytologist who examined stained cells with a phase contrast light microscope, a particular kind of light microscope that enhances contrasts of transparent and colorless objects, making it easier to see the components of a cell. Kellenberger made important contributions to the development of electron microscopes and later to ways of preparing samples for viewing by electron microscopy that produced a minimum of artifacts. Both scientists produced spectacular images.

Bacteriologists were fascinated to see in detail for the first time what the insides of bacteria looked like. I recall attending a seminar when I was a graduate student that was given by Robinow at Berkeley in 1950. He richly illustrated his discussion with slides of exquisite micrographs. The audience was spellbound. When the seminar was over my professor, Roger Stanier, put aside his usual probing postseminar questions in favor of humble praise and requests to see certain slides again.

Electron micrographs, notably those of Kellenberger as well as Robinow, revealed a distinct nuclear area (later called a nucleoid) within bacterial cells that appeared to be evenly filled with fine filaments, presumably strands of DNA. Unlike the true nucleus of eukaryotes, this nucleoid area is not bounded by a membrane. It didn't even have a distinct shape; its contour varied from one cell to another. The structure seemed to have penetrating lobes and to ramify—spread or branch out—through the granular cytoplasm. Later in their lives Robinow and Kellenberger collaborated to write a definitive review of the structure of the bacterial nucleoid.

In 1962 the accumulation of such new information about bacterial cells led Stanier, with the collaboration of his mentor, C. B. van Niel, to publish a now-classic paper, "The Concept of a Bacterium," that lucidly summarizes the properties of the bacterial cell, which were then assumed to be unique. (The archaea, of course, were still unknown.) This paper, with its modest aim of defining what the word "bacteria" meant, was to have lasting impact. It generated a surprisingly vituperative controversy, which, though fading, lives on today.

From the Tree's Roots to Its Branches

The paper demanded attention, both because of the audacity of its claims as well as the prominence of its authors. Van Niel was then at Hopkins Marine Station, a small off-campus laboratory of Stanford University in Monterey Bay, California. He had been recruited from the Technical University in Delft, where he had been a student of the highly admired Albert J. Kluyver. Kluyver was himself a student of the illustrious Martinus Beijerinck, who discovered biological nitrogen fixation and the existence of viruses. Delft itself was where bacteriology began, where in 1676 Antonie van Leeuwenhoek, the "father of microbiology," saw bacteria using his homemade, handheld microscope. In addition to a distinguished background, van Niel was also known for the highly admired summer class in microbiology that he taught at Hopkins Marine Station. A considerable number of mid-twentieth-century microbiologists were introduced to the field there, and his course was justly famous. I took his class in the summer of 1948. A researcher as well as a teacher, van Niel developed the unifying concepts linking the varied types of photosynthesis found in bacteria and in plants.

At the time of Stanier and van Niel's paper, the term "bacterium" was of course well established. Since the late nineteenth century, researchers generally agreed upon which organisms qualified as bacteria, but it remained a loose term. Rather like Justice Potter Stewart's famous comment on obscenity, bacteria were readily recognizable but formally undefined.

Stanier and van Niel's publication changed all that. Stanier introduced the topic in his signature style, crisp yet passionate. The "abiding intellectual scandal of bacteriology has been the absence of a clear concept of a bacterium." After his death, Stanier was criticized, unfairly I believe, with abandoning phylogeny in the case of bacteria. Many cited as evidence this comment, among others, "Evolutionary speculation constitutes a kind of metascience, which has the same fascination for some biologists that metaphysical speculation possessed for some medieval scholastics. It can be considered a relatively harmless habit, like eating peanuts, unless it assumes the form of an obsession; then it becomes a vice."

When Stanier and van Niel's paper was published, before the flood of molecular sequence information became available, speculation about bacterial phylogeny was indeed a metascience. They had not abandoned hope for constructing a phylogenetic taxonomy of bacteria, but they acknowledged that with the information and methods then available, it was not possible. Van Niel, who had a long and abiding interest in bacterial taxonomy, had come to this conclusion much earlier, in 1946. Until more or different kinds of information became available, it was not feasible to develop a phylogenetic classification for these organisms. Van Niel even recommended that Linnaeus's scheme of binomial nomenclature should not be applied to bacteria until their phylogenetic relationships could be discovered. How, he asked, could the required genus component of the binomial be rationally applied without knowing such relationships?

The goals of Stanier and van Niel's paper were thus quite modest. They aimed to set the limits of the group called bacteria. First, they needed to distinguish them from viruses. At the time, some organisms were thought to occupy a middle ground between bacteria and viruses, and some believed that bacteria and viruses might form a biological continuum. At the other limit, they aimed to differentiate bacteria from other small cellular microbes. Their paper was written almost twenty years before the archaea were discovered, so they were not included in the discussion.

Stanier and van Niel dealt swiftly with viruses, repeating a definition written five years earlier in 1957 by André Lwoff in his paper "The Concept of a Virus." Lwoff, a Nobel Prize–winning French microbiologist, proposed that bacteria are cellular organisms and viruses are not. Viruses are acellular, consisting of just an informational macromolecule—RNA or DNA but never both—along with few or in most cases no enzymes, wrapped in a protein coat and sometimes tied in a membrane-enclosed package. This definition prescribed that there could be no intermediates between viruses and bacteria; they are either cellular or they are not. Indeed, when better methods became available, the suspected bacteria-virus intermediates, organisms such as rickettsia, chlamydia, and mycoplasma, were shown to be cellular organisms—bacteria not viruses—even though some of them, as is the case with viruses, are restricted to intracellular growth in their hosts.

Central to bacteria's distinction from other small cellular microbes—the protists—is their unique intracellular architecture that electron micrographs of thin sections had so dramatically revealed. At the time, however, there was no descriptive term in general use to describe bacteria's special cellular organization. Stanier and van Niel realized the need for one. Lwoff, whom they had depended on for setting the limit between bacteria and viruses, suggested that such a term already existed. He referred them to a little-known paper that his mentor Edouard Chatton, a specialist on protozoa, had written in 1937. Chatton used the term "procaryotic" (later spelled with a less Gallic and more appropriate Greek "k"), meaning "before or in lieu of a nucleus," to describe the cellular structure of bacteria. The term "eucaryotic" (later "eukaryotic," meaning "ordinary nucleus") he used for the structure of all other cellular organisms. Bacteria were dubbed prokaryotes; all other organisms, including many microbes, including protists, were eukaryotes. Stanier and van Niel drove this point home: "*The distinctive property of bacteria and blue-green algae (later called cyanobacteria) is the procaryotic nature of their cells.* It is on this basis that they can be clearly segregated from all other protists (namely, other algae[,] protozoa and fungi), which have eukaryotic cells."

The terms were quickly embraced by biologists and have been generally attributed to Stanier and van Niel, who popularized them, even though their originator appears to have been Chatton. Although Stanier and van Niel were careful not to use prokaryote and eukaryote in a taxonomic sense, for they were defining terms, not classifying ones, the terms took on that connotation. Soon most biologists identified prokaryotes and eukaryotes as superkingdoms or urkingdoms, the master divisions of the biological world.

Major complications arose, of course, when the archaea were discovered. They too have prokaryotic cells, but they are not bacteria. Most biologists, by then well conditioned to altering their concepts of taxonomy as new information became available, adapted easily. They simply accepted that there are eukaryotes and two very different kinds of prokaryotes: bacteria and archaea, not closely related but sharing the same distinctive cellular architecture. The terms "prokaryote" and "eukaryote" remained convenient and useful group descriptors outside of the science of taxonomy.

But Woese and his closest admirers, who sometimes identified themselves as "Woese's Angels" or "Woese's Army," reacted quite differently. After Woese's discovery of the archaea, they believed that the very word "prokaryote" was a threat to clear thinking and, of course, to the magnitude of Woese's discovery because it implied a relationship between two completely different groups of organisms, thereby seeming to diminish the importance of archaea. Norman Pace, a leader in the field of molecular microbial ecology, recommended that the word be expunged from polite scientific exchange: "Microbiologists need to take the lead in removing procaryote and similar terminology from textbooks and the lexicon of biology."

Pace and some others remain inexplicably passionate about expunging the term. For example, in 2009 Pace wrote, "The procaryote notion distorts and misleads. Taken literally, it would have stipulated that experiments to probe distinctions between archaea and bacteria would not be necessary. Procaryote smothered interest in microbial evolution and obstructed acceptance of archaea, the first test of the procaryote hypothesis. The procaryote-eucaryote dogma has *paralyzed* (my emphasis) thought and teaching on the origin of the eucaryotic nucleus. It permeates our textbooks and journals with subtle and not-so-subtle misinformation. Thus, the procaryote-eucaryote dogma damaged and continues to damage microbiology by retarding, even denying, progress. It elicits false concepts and misdirects inquiry. There is no place for procaryote in modern scientific discourse." I know of no other example of a term that is considered a threat to scientific progress. Certainly the word is useful. Pace explicitly recognizes the need for it with his use of the awkward substitute, "noneucaryote."

In 1994 Carl Woese had taken a similarly doctrinaire position against the prokaryote-eukaryote distinction, noting its ill effects on a number of scientific endeavors from instruction to research funding. "Can you understand why I have such distaste for the prokaryote-eukaryote dichotomy? This is not the unifying principle that we all once believed it to be. Quite the opposite: it is a wall, not a bridge. Biology has been divided more than united, confused more than enlightened, by it. This prokaryote-eukaryote dogma has closed our minds, retarded microbiology's development, and hindered progress in general. Biological thinking, teaching, experimentation, and funding have all been structured in a false and counterproductive and dichotomous way."

In spite of his advice, "prokaryote" remains a useful and much-used term, even if it no longer describes just one discrete taxonomic unit. Indeed, if there is a culprit, it is nature herself. Although bacteria and archaea are not related taxonomically, they do indeed share the same cell architecture. How could we ignore this compelling fact of nature?

Ernst Mayr, as I discussed in Chapter 4, made a more powerful defense of the term, suggesting that prokaryotes and eukaryotes deserved recognition as the highest divisions of biology.

The Impact of Woese's Methods

Woese continued to be concerned about the impact of terminology on how his newly discovered branch in the Tree of Life was perceived. Perhaps it was some consolation that the methodical approach he used in his discovery of archaea caught on. New technology dramatically decreased the time and effort required to place a newly discovered organism on the Tree of Life.

Woese's method consisted of measuring the difference in the sequence of nucleobases in the small subunit rRNAs of a pair of organisms. This difference was used to determine their relatedness: the greater the number of differences, the more distantly the pair are related. With the methods first available to Woese, this determination was a major challenge. It meant cultivating pure cultures of each of the pair in a medium containing radioactive phosphate (^{32}P) at concentrations barely below the level of lethality, a procedure that would probably not meet modern laboratory safety standards. Next, he needed to isolate the radioactive 16S rRNA from those cultures; incubate the isolated RNA with ribonuclease T_1; separate the resulting RNA fragments sequentially in two dimensions by electrophoresis on a sheet of cellulose using high voltage (also dangerous); expose the sheet to X-ray film to visualize the locations of the fragments; and examine the film on a light box to determine the differences between the two cultures. Clearly, the procedure was time consuming, as well as technically demanding and hazardous. Performing it well required skill and patience. His colleagues said Woese was a master. The procedure worked only on organisms that could be grown in a liquid culture because the RNA had to be made radioactive, thus restricting its use to microbes and cultured animal or plant cells. Modern methods require only a small sample of an organism's cells or tissues, making it readily applicable to all living things.

These newly developed methods soon changed Woese's approach to taxonomy from demanding to routine. The crux of Woese's method was measuring, or in fact estimating, differences between the sequences of nucleobases in the 16S rRNA of pairs of organisms. Woese's original method could detect only some of the differences, those that affected the activity of ribonuclease T_1, but they were enough to yield convincing and startling results. The new method simplified, produced more data, and extended the power and utility of Woese's procedure.

Although sequencing RNA remained awkward and time-consuming, sequencing DNA became simple, rapid, even automated and cheap. Obtaining an appropriate sample of DNA to sequence also became much easier. It was no longer necessary to culture the organism being studied; tiny samples became adequate. An innovative procedure was developed in 1983 by the fascinatingly nonconforming Nobel Prize winner Kary Mullis called PCR (polymerase chain reaction). PCR allows one to designate a particular region in a small sample of DNA and amplify it many times over, producing a sufficient quantity for sequencing. PCR imitates *in vitro* the way DNA is replicated in the cell and repeats it many times. The two strands of the double helix are separated (by heating), starting materials for more DNA along with a necessary enzyme are added, and each of the separated strands is then replicated, thereby doubling the amount of DNA in the sample. By repeating the procedure multiple times, large amounts of DNA can be made. Moreover, such amplification can be restricted to the part of the DNA that one wants to amplify by adding small tags of single-stranded DNA to designate the ends of the region to be amplified. So by using PCR and appropriate tags, ribosome-encoding

DNA from any organism—from a small piece of newly discovered mushroom, for example—can be sequenced. It's not necessary to culture an organism or use high levels of radioactivity to examine its small subunit RNA; a small DNA-containing sample of the organism itself is all that is needed.

The contributions of the powerful combination of PCR and DNA sequencing are not limited to taxonomy. They have impacted almost all aspects of our lives, from medical care to criminal justice to tracing our ancestry. PCR and DNA sequencing are also being used increasingly for studies on microbial ecology, which have led to some spectacular discoveries. Some, probably most, ecologically important microbes cannot be cultured in the laboratory, at least so far. These new techniques make culturing unnecessary for some purposes. One only has to scoop up a sample from the environment of interest, amplify the small subunit RNA-encoding DNA it contains, and sequence it. One of the more startling discoveries using this approach was made by Edward DeLong. He sampled water in the Pacific Ocean off the California coast. The ribosomal subunit-encoding DNA in the sample had the signature of archaea. Further studies showed that such archaea are a major component of these and indeed most temperate ocean waters. This finding came as a major surprise because archaea were thought to be typical of extreme, often thermophilic environments, not temperate oceans. Indeed, by examining the Pacific Ocean's DNA, De Long discovered that it contained a psychrophilic (cold-loving) archaeon, *Cenarchaeum symbiosum*.

In 2004 Craig Venter, a leading player in the sequencing of the human genome, and his colleagues expanded DeLong's approach to microbial ecology through studies on the Sargasso Sea. He and his colleagues took large samples (170 to 200 liters) of water, and by selective filtration harvested the prokaryotic-cell-sized particles from it. They then sequenced all of the DNA the samples contained, a total of 1.045 billion pairs of nucleobases. Comparing these with known sequences of organisms and genes, they identified a number of microbes one might expect to find as well as 148 unknown ones. They also identified known as well as previously unknown genes—over 1.2 million previously unknown, presumably prokaryotic genes. Clearly, we have a lot more to learn about the microbes on our planet.

These powerful new methods have enabled taxonomists to accumulate a vast storehouse of ribosome-encoding DNA sequences, the raw material for constructing and fleshing out the Tree of Life. This remarkable resource is available to all via online databases. One of the most important of these is the Ribosomal Database Project at Michigan State University (http://rdp.cme.msu.edu/), which collects small subunit RNA sequences (16S) of bacteria and archaea. In 2008 it listed 33,082 archaeal and 643,916 bacterial small subunit ribosomal RNA sequences. In 2015 the total was 3,224,600 16S rRNA sequences and 108,901 fungal 28S rRNA sequences. These databases also supply computer software to align and make comparisons among sequences as well as to construct new branches of the tree. Using these technologies, all that is needed to place an organism on the Tree of Life is a small bit of its tissue. It's also possible to locate microbes that have never been cultivated on the Tree of life.

In 2009 Norman Pace accumulated these data to update the Woese-based Tree of Life (see Figure 3). The greatest improvement was seen in the bacterial branch. Woese had identified twelve main phyla; Pace could now identify seventy, only half of which had representatives that had been cultured. Like earlier representations of the tree, there was little branching among the bacteria. The various lines of progression all appeared to spring simultaneously from a common branch, a "basal radiation" representing a sudden and explosive diversification of species. In contrast, the archaeal branch did not expand correspondingly. Pace divided the archaeal domain into two phyla, the Crenarchaeota and the Euryarchaeota, as Woese had done.

In the years following the publication of Pace's tree, technology has expanded our ability to sequence even single cells. In addition, we are now better able to determine the sequence of individual cells in a natural population by sequencing the aggregate DNA in a particular environment. In 2016 a group of seventeen scientists from nine institutions led by Jillian F. Banfield at the University of California, Berkeley, used much of this new information to construct an updated version of the Tree of Life, which was published in the journal *Nature Microbiology*. The popular science writer Carl Zimmer announced the feat on April 11, 2016, on the front page of the *New York Times*. Banfield's team assembled information from 2,072 genomes that are now publicly available along with 1,011 reconstructed genomes that they have recovered from a variety of environments, making a total of 3,083 genomes, to present what they called in *Nature Microbiology* "a new view of the tree of life." As in previous studies, Banfield and her colleagues used the ribosome as the source of their comparisons, but rather than using the sequence of the small subunit RNAs, they utilized the sequences of sixteen different ribosomal proteins from each organism. They employed a supercomputer to assemble this vast quantity of data into a tree. The results were startling. Archaea were shown as only slightly more diverse, but bacteria were many times more diverse than before. The most dramatic development was the addition of a totally new branch of the bacterial domain that they termed "candidate phyla radiation" (CPR). This branch is termed "candidate" because no representative of this huge assemblage has yet been cultivated. CPR are estimated to be about as diverse as all the other bacteria combined. Members of the CPR are small organisms, perhaps endosymbionts or a primitive group of the domain. Certainly this remarkable addition to the

Tree of Life reminds us how much we have yet to learn about our microbial relatives.

The Tree of Life may have a firm molecular basis, but like all things biological, our understanding of it is always evolving and open to debate.

{ PART TWO }

DOUBTS AND COMPLICATIONS

CHAPTER 6

Genes from Neighbors

Among the many statements that can be made by analyzing the Tree of Life is a seemingly uncontroversial one, that organisms derive their inheritance—their genes—from their parents. In the case of prokaryotes that multiply asexually, they inherit from the mother cell that divided to form them. The tree implicitly insists that inheritance is vertical: genes are passed down from one generation to the next, parent to offspring. Constructing a Tree of Life or even a family tree from any source of data is similarly predicated on an assumption of vertical inheritance. Although this seems like a safe assumption, there are complications, especially in the case of prokaryotes. In their realms, one often encounters horizontal or lateral inheritance, the transfer of genes from a neighbor instead of from a parent.

DNA on the Loose

Vertical inheritance was first questioned not quite a hundred years ago in the laboratory of a shy British pathologist and microbiologist, W. Frederick Griffith, who was working in a somewhat antiquated laboratory of the Ministry of Health in Dudley House on Endell Street, London, which he shared with a close colleague, William Scott. Years later, on April 17, 1945, Griffith and Scott both died when the apartment they shared took a direct hit in the blitz of London during World War II. Their Endell Street laboratory, situated above a post office, was isolated and ill equipped, according to visitors. Nevertheless, paradigm-changing discoveries about gene transfer among bacteria and the mechanism of inheritance occurred there. These discoveries, published in 1928, also led directly to the discovery of "idioplasm" (a wonderfully appropriate, now archaic word meaning roughly "the substance that makes us what we are"), the self-replicating material on which inheritance is based and from which genes are made, which we now know as DNA. But in Griffith's time DNA, in fact any kind of nucleic acid, wasn't even a minor candidate for encoding inheritance. It was judged to be far too simple a compound to play such a critical, information-packed role in biology. The much more complex protein molecule seemed more suited to the task. Nucleic acids (DNA and RNA) are composed of only four different kinds of component nucleobases. Proteins are composed of twenty different kinds of component amino acids, seemingly offering far more potential.

Griffith, at the time of his momentous discovery, was engaged in a topic quite different from horizontal inheritance or how inheritance was encoded. He was immersed in the study of the epidemiology of infectious disease: Where do the pathogens hide that cause disease, and how do they spread through a population? How his path diverged from disease to DNA and an understanding of horizontal gene transfer is an example of the serendipities of the scientific process and a story well worth recalling.

Griffith realized that he had to be able to recognize and distinguish strains of a bacterial pathogen that appeared to be superficially identical if he were to trace the source and progress of an outbreak of disease or even control it. He focused his attention on *Streptococcus pneumoniae,* commonly called pneumococcus because of the shape of its cells: its spheres (cocci) characteristically occur in pairs that cause pneumonia. The particular pneumonia that pneumococcus causes is called lobar pneumonia because it affects an entire lobe of the lung (sometimes both lobes, in which case it is called double pneumonia). In Griffith's time, lobar pneumonia was deadly, before it

Genes from Neighbors

could be cured with antibiotics and prevented by vaccination. About 30 percent of untreated patients died. Untreated lobar pneumonia can kill susceptible, healthy adults in a matter of days. I well recall a robust, close colleague of my father who died of lobar pneumonia before we even knew he was ill. Pneumococcus is also a common cause of ear infections of children, which, though sometimes extremely painful, are usually self-limiting, not lethal.

Untreated patients who survived pneumococcal pneumonia typically underwent a crisis six to ten days into their illness, the event often depicted in old movies when the kindly attending bedside physician stands and announces with a sigh of deep relief that the loved one will survive; the flickering bedside candle again burns brightly. The patient experienced drenching sweat and a sudden disappearance of fever. The crisis occurred when sufficient time had transpired for the patient to make antibodies against the invading pneumococci, allowing the patient's phagocytes to destroy them.

In order to recognize and identify individual strains (isolates) of pneumococcus and thereby trace their lineage in an outbreak, Griffith set about "typing" them according to their immune reaction in mice. This process is known as serology because blood serum (the noncellular fraction of blood and the major repository of an organism's array of antibodies) is used to test the reaction. If a strain of pneumococcus, that is, a particular isolate from a patient suffering lobar pneumonia or some other pneumococcal infection, can be inactivated (caused to aggregate into visible clumps of cells) by serum from a mouse that had been exposed to another isolate of pneumococcus, the two strains are said to be the same type. By testing a large number of isolates from various patients, Griffith found that the same set of types, designated I, II, III, and IV (the latter was subsequently divided into many more types), of pneumococci in patients in London hospitals were the same that scientists in New York had reported finding in theirs. Types of pneumococci are apparently international, an encouraging result for Griffith. Perhaps by typing, the worldwide spread of pneumococcal outbreaks could be traced.

But not all strains of pneumococcus proved to be typeable by these means. Griffith found these untypeable strains in some patients that were recovering from a pneumococcal infection. Such strains proved to be either weakly virulent or completely innocuous forms of pneumococci. They could not be typed because they didn't stimulate formation of immunologically active serum or react to it. All such benign strains shared an easily recognizable characteristic (phenotype): their cells all lacked the thick outer layer of gummy polysaccharide material, called a capsule, which surrounds, confers disease-causing ability (virulence), determines type, and typifies pneumococci. This bulky capsule plays a critical role in infection. It acts as armor for an invading pneumococcal cell by preventing the host's protective leucocytes from engulfing and killing it. Its disease-causing mechanism is simple: the capsule makes the pneumococcal cells too large and charged for a leucocyte to swallow. Protective antibodies such as those accumulating in the patient undergoing a crisis form a bridge between the leucocyte and the pneumococcal cell, allowing the leucocyte to engulf and kill the pneumococcal cell.

Colonies of untypeable, nonvirulent strains are readily recognizable when grown on Petri plates because colonies of them have a different, less glistening appearance. Griffith dubbed the colonies and strains that produced them rough (R). In contrast, the shiny colonies of all types of capsule-forming, virulent strains of pneumococci were called smooth (S). Suspecting that the presence of such R strains might play an active role in determining whether a patient survived the infection, Griffith focused on the origin of these benign R strains. Could he cause them to appear and thereby possibly cure an infection? He soon learned how to generate at will R strains in vitro by mimicking the environment of a recovering pneumonia patient. If he grew various types of S strains in the presence of their inhibitory, type-specific serum, R strains would emerge and dominate the population. Clearly he was selecting, not generating, R strains: spontaneous mutations constantly give rise to R cells, and in the presence of S-cell-suppressing antiserum, only they survive to become the majority.

Curious as to whether R strains could regain virulence and thereby reinitiate an infection, Griffith injected large numbers of R cells into mice. The mice died. He found that they were killed by S strains of their original type, that is, the type from which the R strain had been isolated. The R strain had quite clearly reverted by mutation to its original S state, an event that could be lethal for a recovering patient. Griffith, again with an eye toward treatment, then set about investigating the factors that might influence such reversion from nonvirulent R back to its original, virulent S state. He suspected that formation of what he observed and called a "nidus" or accumulation of clumps of cells was a prerequisite to such reversion. To stimulate formation of a nidus, he injected heat-killed S cells to form a nidus along with the live R cells. His hunch, though not its scientific underpinnings, was correct: the presence of the heat-killed S cells did indeed stimulate the appearance of live S cells.

Griffith did many such experiments, including one that led to the remarkable discovery for which he is remembered. Along with R cells, he injected heat-killed cells of an S strain of a different type (I) from the original type (II) from which the R strain was derived. Again the mice died, and again S strain cells could be isolated from the dead mice, but they were type I. The R cells hadn't reverted to their original type II; they had been "transformed"—presumably by the presence of dead S cells—into becoming a different type, that of the dead S cells. Griffith repeated this experiment with a number of variations, confirming and reconfirming the observation: dead S cells are indeed able to confer their type status on live R cells, transforming them. Although the mechanism was mysterious, the consequences were clear. The R cell had inherited the characteristic (of being type I) not from its parental mother cell but from a neighboring cell. Griffith had shown not a conventional case of vertical inheritance from a parent but one of lateral gene transfer or horizontal inheritance from a neighbor, albeit a dead one.

Griffith apparently found these results as disturbing as they were startling, as most of us probably would have. They didn't make sense, but he couldn't deny them. Repeated experiments yielded the same results. As he somewhat reluctantly conceded, "there seems to be no alternative to the hypothesis of transformation of type." For his epidemiological purposes, he would have preferred that a strain's type status be immutable and therefore traceable. He was slow, reluctant, and cautious about publishing his results. Only after a visiting colleague, Friedrich Neufeld, had confirmed Griffith's results in his own laboratory in Berlin did Griffith agree to publish his findings. He did so in 1928 in an exceptionally long (forty-seven pages) and detailed paper. The paper convincingly established that a mechanism of horizontal gene transfer that came to be known as transformation occurred among pneumococci. That was Griffith's last brush with pneumococcus. He had had enough of it. In his remaining years he investigated the epidemiology of other bacterial pathogens.

Studies on the mysterious phenomenon of transformation, however, continued. It caught the eye of Oswald Avery, another shy physician and microbiologist, a Canadian working at the Rockefeller University Hospital in New York City. Avery repeated Griffith's experiments, confirmed them, and pursued the fascinating phenomenon in his own larger, well-equipped laboratory. There, rapid advances were made that led to a series of discoveries. First, Avery's team discovered that such transformations do not have to occur in a mouse; they can take place *in vitro*. Merely growing R cells in a test tube in the presence of heat-killed S cells can transform the original type of some of the R cells. Then, researchers in Avery's laboratory found that intact cells weren't required for transformation to occur. Extracts made from S cells could transform the type of R cells. Something in the extract, which he called the "transforming principle," caused the genetic change. Of course, at the time of these studies, the chemistry of inheritance was unknown. To identify the material in which genetic information is stored, the elusive idioplasm, Avery and his team proceeded to purify the transforming principle. The purified material proved to be DNA! Even Avery was surprised. He wrote his brother, Roy, that the transforming agent "is in all probability DNA: who could have guessed it?"

Avery, along with two colleagues, Colin MacLeod and Maclyn McCarty, published their startling results in 1944. For a number of reasons, many in the scientific community, including its opinion-forming leaders, were reluctant to believe that DNA was the molecule in which genetic information was encoded. Perhaps, it was suggested by the doubters, Avery's highly purified DNA was contaminated with small amounts of protein that actually mediated transformation, even though Avery had shown that addition of the DNA-destroying enzyme deoxyribonuclease (DNase) obliterated the transforming activity of his preparation. DNA's composition was thought to be too simple. A harsher reason existed for the dismissal of Avery, MacLeod, and McCarty's paper: Avery was not a member of the developing and somewhat exclusive molecular biology club that focused on the study of bacteriophages. Avery was a medical microbiologist, not a molecular geneticist. He published his results in a medical journal, not a genetic or biochemical one. He had a retiring personality in stark contrast to the self-assured flamboyance typical of many of those who studied molecular biology.

It was a much simpler and more highly equivocal experiment, the so-called Hershey-Chase experiment completed eight years later, that proved attractive and convincing to the molecular genetics community. Perhaps Avery's experiments seemed dogged: Hershey and Chase's had flair. Their 1952 experiment is frequently cited even today as the one that identified DNA as genetic material (idioplasm).

Alfred Hershey and Martha Chase's experiments used a bacterial virus called T2 phage (a darling of molecular geneticists) that infects certain strains of *E. coli*. For a virus, T2 phage has quite a complex structure: an icosahedral head is attached to a cylindrical tail with some fibers and a plate at its end. Electron micrographs show that when phage T2 infects a bacterial cell, the plate attaches to the cell surface, and the phage particle itself, although its head appears to be empty, remains on the surface until the bacterial cell bursts as a consequence of the infection. The phage and its action are somewhat reminiscent of a hypodermic needle with an attached bulb. When a phage particle (virion) attaches to a susceptible bacterium, its tail shortens, and it injects the content of its head into the bacterial cell. Chemical analysis had already established that phage T2 is composed exclusively of protein and DNA. One or the other of these components had to be the phage's genetic material that directed the formation of more phage particles. That's all there is.

To identify which component carried the virus's genetic instructions, Hershey and Chase differentially tagged the protein and DNA components of phage T2 so that their subsequent fates could be followed during infection. They labeled the protein with radioactive sulfur (^{35}S) and the DNA with radioactive phosphorus (^{32}P). (DNA does not contain sulfur and protein does not contain phosphorus.) Soon after infection, at a time when they knew that the phage heads

Genes from Neighbors

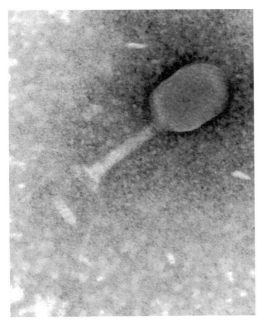

FIGURE 10. Electron micrograph of bacteriophage T4, which is almost identical in appearance to bacteriophage T2, the phage used in the Hershey and Chase experiment to prove that DNA is the genetic material. The icosahedral "head" is packed with DNA. The rest of the structure is made of protein.

would be empty, having injected their contents into the cell, but before the infection had proceeded to a point when infected cells would begin to burst (lyse), they put the phage-bacterial cell mixture into a Waring blender (the band leader Fred Waring's unanticipated contribution to molecular genetics) to shear phages with their empty heads from the infected bacterial cell. They then centrifuged the mixture at a low speed sufficient to spin the bacterial cells to the bottom of the tube while leaving the empty phage parts in the fluid at the top of the tube. When DNA was labeled, the radioactivity went with the cells. When protein was labeled, it did not. Its radioactive label was found principally in the fluid with the empty phage parts. From this they concluded that the phage's DNA, not its protein, must have carried the information into the cell that caused it to be infected and destroyed. So, they concluded, DNA must be the genetic material.

The Nobel Prize committee took note of the Hershey-Chase experiment, and in 1969 Alfred Hershey shared the Nobel Prize with Max Delbrück and Salvador Luria. Neither Fredrick Griffith nor Oswald Avery, who had published the same conclusion eight years earlier, was awarded a Nobel Prize. Clearly, being a shy, soft-spoken, medical microbiologist doesn't pay.

Although Griffith's studies are best remembered for having led to the momentous discovery that DNA encodes genetic information, they also constitute the first revelation that genes can be transferred horizontally as well as vertically, at least among some prokaryotes. The mechanism Griffith discovered that came to be known by the curious term *transformation* is one that reminds us of the way fish procreate: one prokaryotic cell releases DNA into its surroundings and another cell takes it up and integrates it into its own genome, thereby becoming genetically altered or "transformed." (My mother used the term as a euphemism for a lady's wig.) The status and ambiguity of the word is now augmented by its use in a completely different way—to describe the transition of a cell from a normal to a cancerous state. Yet the term is still used to describe this particular kind of horizontal gene transfer.

A cursory consideration of transformation among prokaryotes engenders images of some sort of genetic orgy: genes being indiscriminately spewed from some cells and randomly taken up by others, but such is certainly not the case. Transformation as it occurs naturally among prokaryotes is a highly ordered, genetically controlled process. Only a minority of prokaryotes are capable of this type of genetic exchange, and those that are follow a tightly scripted scenario. Even in species such as the pneumococcus *(Streptococcus pneumoniae)*, which as Griffith had discovered readily undergo transformation, cells are not always able to take up DNA from their environment. They must first enter a susceptible physiological state called competence. The process of becoming competent and successfully taking up DNA from their environment is under strict genetic oversight. In excess of twenty genes of the pneumococcus are dedicated to the processes of becoming competent and undergoing transformation.

Another naturally transformable bacterium, *Haemophilus influenzae,* is even more restrictive. Not only must it become competent to be transformed, the right kind of DNA must be available in its environment. *H. influenzae* takes up only DNA that has been released by other *H. influenzae* cells. It recognizes this DNA as coming from a fellow *H. influenzae* cell by the presence of stretches of specific ten-base pair sequences that are scattered throughout all *H. influenzae* DNA. At least in some cases, the release of DNA from donor cells is also regulated: evidence suggests that cells of the naturally transformable species *Pseudomonas stutzeri* release DNA into their surroundings only when they come in physical contact with another *P. stutzeri* cell.

Certainly, such prokaryotes that are genetically able to undergo transformation are likewise capable of horizontal gene transfer, but it's quite restricted. Very few prokaryote species are endowed with a capacity to undergo transformation naturally, and these few are capable of transferring genes only to and from cells almost identical to themselves. The selective advantage of such transfer is hard to fathom, but there has to be some reason for it to have evolved because those bacteria that do participate pay a nontrivial genetic price. Natural transformation requires the evolution of a significant number of genes dedicated to this capability. It would seem to play only a minor role in the vast sea of genetic exchange among organisms. However, natural transformation, that is, genetically programmed transformation, is not the only way that cells can be transformed by incorporating DNA that might be present in their environment.

In 1970 two microbiologists, Manley Mandel and Akiko Higa at the University of Hawaii, changed profoundly our concepts of the quantitative impact of transformation. They found that natural transformation is not the only way that DNA from the environment can enter a bacterial cell. In a four-page letter to the editor of the *Journal of Molecular Biology*, they reported that if cells of *E. coli* (a species not endowed with a natural capacity to undergo transformation) are treated with calcium ions (Ca^{++}), a huge DNA molecule constituting the complete genome of the bacterial virus, lambda, can readily enter its cells, a discovery that ushered in the era of recombinant DNA technology and changed fundamentally our understanding of the impact of horizontal gene transfer by transformation.

Two years later at Stanford University, Stanley Cohen, Annie Chang, and Leslie Hsu extended these observations on artificially induced transformation and improved the efficiency of the method. They found that calcium treatment followed by an abrupt increase in temperature renders *E. coli* able to take up any sort of plasmid DNA.

Now by a variety of treatments, including exposure to a pulsating electrical field (electroporation), it's possible to transform almost any kind of cell artificially, prokaryotic or eukaryotic. Indeed, such artificial transformation is at the core of recombinant DNA technology, the basis for the biotechnology industry and much of modern biology. It allows DNA that has been manipulated *in vitro*, for example, in a process that joins pieces of DNA from different sources into a single molecule, to be introduced into a living cell, where it can be expressed and replicated. All recombinant DNA technology, including cloning and production of medically important human proteins such as insulin and human growth hormone, depends on artificial transformation.

At first glance artificial transformation, dependent as it is on harsh laboratory treatments, would appear in comparison to natural transformation to be artifactual and irrelevant to microbial evolution in a natural environment. But quite the opposite is true, or at least closer to the truth. As we've seen, natural transformation is limited to a minority of prokaryotes, and it is elaborately choreographed to restrict genetic transfer to exchanges between closely related strains. In contrast, artificial transformation can be induced to occur in virtually all microbes, prokaryotic and eukaryotic, and there are almost no restrictions on the relatedness of DNA that can be transferred. DNA from humans can be transferred to *E. coli*, for example.

The sorts of laboratory procedures that make artificial transformation possible are not those one would expect to encounter in nature, but their effects on the cell might not be restricted to laboratory manipulations. As disparate as they are, the treatments employed to mediate artificial transformation share a common feature: They damage a cell's surrounding envelope of membrane and wall in such a way that DNA can pass through it, but the damage to the membrane can be repaired later. Natural environments must be replete with cell envelope–damaging conditions and chemicals, some of which would allow transient access to environmental DNA. And environmental DNA is likely to be present in areas thickly populated with microbes: many kinds of cell death lead to spilling of a cell's DNA into its environment. Perhaps artificial transformation is not that artificial. It might occur relatively frequently in natural environments, particularly in those with tightly packed microbial populations such as biofilms. In these locations, the sheer density of life may lead to massive horizontal gene transfer.

Microbial Sex

Transformation is not the only known mechanism of horizontal transfer of genes among prokaryotes. Some eighteen years after Griffith announced his discovery of horizontal gene transfer by transformation, a second mechanism of lateral transfer was discovered, in this case as a result of a deliberate search for such transfer. The discoverer, Joshua Lederberg, was a scientific prodigy who graduated at fifteen from New York's Stuyvesant High School for science and technology. He was inspired at a young age by reading Paul de Kruif's book *The Microbe Hunters,* which he said "turned my entire generation toward a career in medical research," and enrolled as a zoology major at Columbia University in preparation for medical school.

At Columbia Lederberg met a charismatic professor of biochemistry, Francis J. Ryan, an encounter that was to change his life course abruptly. Lederberg, anxious to get an early start on his dream to "bring the power of chemical analysis to the secrets of life" asked to do research with Ryan, who was excited about Avery's recent discovery that the transforming principle (and by extension the discovery that the repository of genetic information) was encoded in the structure of DNA. Ryan at the time was engaged in the study of the red bread mold *Neurospora crassa,* which had come to be a favorite of those pursuing the genetic encoding of metabolism. Two Stanford University professors, George W. Beadle and Edward Tatum, had popularized use of this fungus in their investigations on the consequences of mutations on the biosynthesis of certain small molecules, including amino acids. They had shown that a single mutation could inactivate an enzyme and that some of such inactivations could render a strain of *N. crassa* incapable of synthesizing a particular amino acid. The mutation changed a gene such that the enzyme it encoded was no longer active and thereby unable to catalyze a vital step in the synthesis of that amino acid. The mutant strain could grow only if that amino acid were made available to it from the environment. Beadle and Tatum's studies led to a unifying principle that was summarized as "one gene, one enzyme." Genes encode enzymes and enzymes do the cell's work. Although this dictum was shown later to have exceptions (two genes are needed to encode certain enzymes), it brought with it an exhilarating sense of logic to the subject, linking genetics and metabolism, the beginnings of the field of biochemical genetics. I recall the wonderful feeling of clarity and the insight I experienced on first reading those papers. At the time I even focused my graduate work on the water mold *Allomyces macrogynus* in an attempt to make it a useful tool for extending such investigations.

Because Ryan, Lederberg's mentor, had recently spent an academic leave at Stanford with Beadle and Tatum and had brought cultures of *Neurospora crassa* with him, the nexus of people (Ryan and Lederberg), ideas (transformation and biochemical genetics), and available, amenable biological tools (*N. crassa* and biochemical mutant strains of it) seemed almost to demand an answer to this question: Can cultures of *N. crassa* be transformed? Their experiments indicated that the answer was a clear "no," at least not by the methods they employed, but the intellectual fuse had been lit.

Next Lederberg turned his attention to seeking evidence of genetic exchange in another microorganism that was beginning to be studied genetically the way *Neurospora* had, namely the bacterium *Escherichia coli*, then an ingénue in the world of molecular biology. Again the results were negative. Not to be deterred, Lederberg shared more ideas and approaches to investigating genetic exchange among bacteria. Ryan agreed that the question should be pursued, but suggested that Lederberg continue such studies with his former Stanford colleague, Edward Tatum, now not too far away at Yale University. Tatum had a collection of mutant strains of

Escherichia coli and more experience with this experimentally amenable bacterium.

Lederberg was then in the V-12 Navy College Training Program at Columbia University and was expected to complete his undergraduate medical program quickly and serve in the armed forces during World War II. He asked for and was granted a year's leave to continue his genetic studies on *E. coli* with Tatum. The two of them were spectacularly successful.

Tatum had a collection of mutant strains of *E. coli* at Yale that he had generated by exposing the cells of this bacterium to X-rays or ultraviolet light. These strains, like those of *N. crassa* that Beadle and Tatum had studied, belonged to the class called biochemical mutants, that is, they were biochemically impaired in their ability to synthesize an essential nutrient on their own. As a consequence of such impairment, they could grow and reproduce only if that particular nutrient were supplied to them from their environment (present in their culture medium). Such growth/no-growth experiments allowed researchers to see readily whether a strain of the bacterium did or did not carry a particular mutation.

Tatum and Lederberg were looking for evidence that genes in one cell could mix with those of another, resulting in recombinant cells that had either both or neither of their parents' abilities to make an amino acid. With the availability of these mutant strains, the critical experiment to test for genetic exchange seemed straightforward: mix two different mutant strains—for example, one that needed the amino acid histidine (which they had designated H^-), and one that needed the amino acid methionine (M^-)—hope that they might exchange genes, and then look for a recombinant strain, that is, a strain that required both histidine and methionine or one that required neither. It was easy enough to test for possible recombinants that needed neither of the amino acids to grow: simply spread the mating mixture on the surface of a Petri dish containing an agar-gelled culture medium that lacks both amino acid nutrients, a so-called minimal medium. The presence of any cells in the mating mixture that could grow on the medium to produce a visible colony would be presumptive evidence of recombination: they grew in the absence of either of the amino acids that one or the other of their parents required.

But Lederberg realized there were complications to such seemingly simple experiments. He and Ryan had already investigated the frequency of reversion of such biochemical mutants, that is, the frequency of occurrence of spontaneous mutations that corrected such nutritional deficiencies, thereby conferring independence from need of nutritional supplementation without necessity of genetic exchange. Lederberg, working with Ryan, had previously found that such reversions are somewhat rare, occurring at a frequency of about one in ten million (10^{-7}), but owing to the huge numbers of cells to be used in their planned experiments, frequent enough, perhaps, to obscure actual rare recombination events.

So Lederberg and Tatum resorted to using double-mutant strains: those that required two nutritional supplements as a consequence of two separate mutations. Their frequency of reversion to independence of nutritional supplements would be expected to be the product of the frequencies of reversion of each individual biochemical mutation, or 10^{-14} ($10^{-7} \times 10^{-7}$), far less frequent than could be detected experimentally. At a frequency of 10^{-7} one would expect on average to find a cell that had reverted in each one-hundredth of a milliliter of a full-grown laboratory culture of *E. coli*. At 10^{-14} one would have to search through a hundred thousand liters of such a culture.

Lederberg mixed a pair of such double-mutant strains: one required the vitamin biotin, designated B^-, and the amino acid methionine, designated M^-; the other required a pair of amino acids, proline, P^-, and threonine, T^-. When mixed, cells with no nutritional requirement were found at a frequency of about 10^{-7} (ten million times more frequently than the reversion rate). In other words, when Lederberg added a billion cells from the mating mixture to a Petri dish lacking nutritional supplements, he found about 100 recombinant colonies—those developing from cells that required none of the four parental nutritional needs.

This simple experiment offered solid proof that cells of *E. coli* could exchange genes, but many questions remained. Perhaps the first to occur to those who read Lederberg and Tatum's succinct, eleven-page page paper was this: How did the *E. coli* cells exchange genes? Was it another example of transformation? The authors didn't deal with this sort of detail. They merely pointed out that recombination and segregation had occurred. We'll discuss segregation later, and why Lederberg and Tatum were wrong about segregation having taken place. Lederberg and Tatum were willing to propose that hybridization had occurred, but had not explained how it had occurred. Later, Lederberg showed that cell-free filtrates, in which one might expect to find transforming DNA if it were present, were incapable of generating recombinants, thus providing circumstantial evidence against transformation being the cause of the hybridization they had observed. Did that imply that bacterial cells had to come into direct sexual contact to effect genetic exchange?

Three years later Bernard Davis, who made many important contributions to microbiology, in what later became known as the bundling-board experiment, answered the question directly, simply, and convincingly. He grew two of Lederberg's cultures, each in one of the arms of a U-shaped tube, separated at the bottom by an "ultrafine" fritted (finely porous) glass disk through which liquid but not cells could pass. As the cultures grew, Davis applied a vacuum alternately to the two arms of the tube, thereby flushing culture liquid but not cells back and forth between the arms. When growth had ceased, he found no recombinants in either arm of the tubes. There was no evidence that genetic exchange had occurred. It seemed clear that genetic exchange depended on cells being able to touch each other, which the "bundling-board" disk had prevented. The genetic exchange Lederberg had found to occur in *E. coli* could not be another example of transformation.

The confounding question of why Lederberg was able to find genetic exchange among cultures of *E. coli* in Tatum's laboratory but not in Ryan's opened new vistas in microbiology. Lederberg's experiments were repeated with cultures from the two laboratories with the same results: negative with Ryan's cultures, positive with Tatum's. Other than location, there was one major difference. The two laboratories used different strains of *E. coli*. Ryan had used a strain designated B, which had been isolated in 1918 at the Pasteur Institute in Paris, and Tatum had used one designated K-12, which had been isolated in 1922 at Stanford University and kept there in their culture collection. Further experiments established that K-12 strains were fertile, but B strains were not. It's a curious fact that the success of Lederberg's Nobel Prize–winning experiments hung from the slim thread of his co-researcher, Tatum, having transported his Stanford laboratory strain of *E. coli* (K-12) to Yale. At the time, strain B was the more popular strain. It is more vigorous. It was the strain used by Max Delbrück's "Phage Group," the one used by Hershey and Chase, and one upon which the burgeoning field of molecular biology depended. Nevertheless, how could two strains of the same bacterium, albeit isolated from different sources, differ in such a fundamental respect?

Lederberg assumed that the cellular basis for the bacterial sex he had discovered was much like sex in other creatures, other than not being required for reproduction. Two haploid cells merge, forming a diploid zygote, which then segregates to produce new combinations of the genes contributed by each of the two parents. That's why Lederberg mentioned that he had found an example of segregation, the principle that offspring acquire a different array of such factors from those found in either parent. His hypothesis fit with the accumulated experience of biology and also with what was then known about *E. coli* sex: bacteria, at least strains of *E. coli*, are known to be haploid. Lederberg knew that they were haploid because clearly recessive mutations, those that repressed ability to perform a task, such as synthesize an amino acid, were expressed. They wouldn't be expressed if the strains were diploid; the sister unmutated gene would carry out the function, thereby masking the mutant gene's loss of activity.

At that point the saga of bacterial sex crossed the Atlantic to the laboratory of William Hayes, an Irishman with a medical and a science degree. At the Hammersmith Hospital of the University of London Postgraduate Medical School, Hayes performed a series of *E. coli* crosses, using the same K-12 strains that Lederberg had used. In January 1952 he published his results in a note of less than one page in the British journal *Nature*. His note drew little attention, although its title, "Recombination in *Bact. coli* K12: Unidirectional Transfer of Genetic Material" announced a dramatic difference between bacterial and other known kinds of conjugation in which both parents contribute equally to a zygote, a total merging of genomes, not a transfer from one to the other. Hayes was proposing that bacterial sex didn't involve the formation of a zygote, but rather the formation of partial zygote or "merozygote" (*mero-* meaning "partial"). The entire genomes of the mating pair did not mix to form a zygote. Rather, one cell passed only certain of its genes to its mating partner—a radical concept. Only in September, when Hayes presented his results at the Second International Symposium on Microbial Genetics in Pallanza, Italy, was their significance fully appreciated. In his book *Double Helix*, James Watson referred to that presentation as a bombshell: "everyone in the audience knew that a bombshell had exploded in the world of Joshua Lederberg." Hayes later modestly accounted for his success by stating that he "had the great advantage of knowing virtually no genetics while Lederberg knew too much!"

Hayes's paradigm-altering experiments were simple enough. He merely added an additional genetic trait—resistance to the antibiotic streptomycin—to one of Lederberg's two strains so he could follow the individual fates of the two strains during a cross. Adding streptomycin would eliminate further involvement of the sensitive strain, rather like shooting one partner of a mating pair. When he crossed two strains, designated 1 and 2, in the presence of lethal concentrations of streptomycin, he obtained recombinants if strain 1 were resistant to the lethal action of streptomycin and strain 2 was killed by it, but not if the susceptibilities of the two to the lethal consequences of streptomycin were reversed. Obviously the crosses exhibited a polarity. Only one member of the mating pair had to survive to produce recombinant progeny, , not unlike reproduction in eukaryotes, in which the female but not the male must survive mating for progeny to be produced.

Hayes concluded correctly, as further research established, that strain 2 was a donor of genetic information and strain 2 was the recipient. The two cells did not fuse. Only the recipient strain 2 had to survive exposure to streptomycin for recombinants to be produced. Strain 1's role was complete once its genes had been donated. Further studies showed that even a streptomycin-killed donor strain one could pass its genes to the recipient, reminiscent perhaps of the male preying mantis, which can carry on with a successful mating even without its head. Strain 1 and like strains that need not survive the mating became known as fertile or F^+, and by analogy to higher forms, also known as male. Strain 2 and like strains, which had to survive a mating to be successful and were unable to act as donors, became known as F^- or female.

DOUBTS AND COMPLICATIONS

Soon more information from Hayes's laboratory and others about this curious form of one-way sex began to accumulate:

- F^+ strains are relatively rare in nature. One study found that of 140 different isolates of *E. coli*, only 9 (6 percent) were F^+. Clearly, Lederberg was quite fortunate that Tatum's K-12 strains just happened to belong to the fertile 6 percent, which Ryan's B strain and most others do not.
- F^+ strains occasionally convert to being F^-.
- Perhaps most surprisingly, the overwhelming majority of products of an $F^+ \times F^-$ cross are F^+. Maleness in the case of *E. coli* and, as it was later discovered, in many other bacteria is genetically transferable, seemingly infectious.

These data fit with a hypothesis that was later proven correct, namely that bacterial maleness is conferred by the presence of a genetic factor that has the property of being able to transfer itself to cells that lack it. Occasionally, as this factor (F) transfers itself to another cell, it brings some of the donor cell's genes along with it.

F, which was later found to be an autonomous bit of circular DNA called a plasmid or an episome, is a quintessentially selfish set of genes. The episome reproduces itself within the cell that carries it and has the ability to spread by transferring itself to other cells, where it reproduces and can then spread further. Rarely, the episome becomes integrated into a donor cell's chromosome. When it transfers itself to another cell, an F^- cell, it drags a bit of the donor cell's chromosome and the genes it contains along with itself. That scenario explains the matings and recombination that Lederberg had discovered. It also explains the difference between strains K-12 and B. Strain K-12 is F^+; Strain B is F^-.

The process by which F transfers itself to another cell is surprisingly elaborate. F encodes a sex pilus, which as a consequence all

Genes from Neighbors

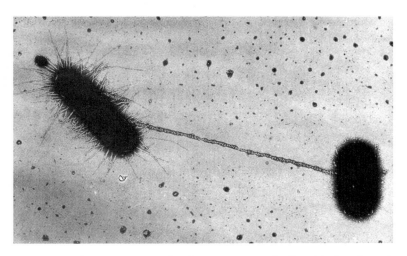

FIGURE 11. Electron micrograph of a mating pair of cells of *Escherichia coli* joined by an F pilus.

F⁺ cells bear. This pilus is a long (several times longer than the cell itself), thin protuberance that extends from the cell surface. When the tip of the pilus encounters an F⁻ cell (it ignores other F⁺ cells), it attaches to the potential recipient. Then the sex pilus, more like a grasping arm than a penis, retracts back into the F⁺ cell, pulling the captured F⁻ cell along as it does. When the two cells come into intimate contact, the F plasmid replicates itself by an odd mechanism called a rolling circle: the circular plasmid replicates one of its strands as it turns, producing a single-stranded DNA molecule that penetrates into the F⁻ cell. Inside its new home, the single-stranded DNA is duplicated and recircularized. The former F⁻ female cell becomes F⁺ and therefore male and fertile.

Occasionally, DNA of the F plasmid becomes integrated into the cell's chromosome. Then in a subsequent mating, it drags chromosomal genes with it into the F⁻, as Lederberg and Tatum had observed. This, also, explains why in matings transfer of the F plasmid is the rule and transfer of chromosomal genes quite rare, one in ten million as Lederberg and Tatum found. When the essence of these mating mechanisms became understood, bacterial geneticists isolated special strains that mediated transfer of chromosomal DNA at a higher frequency, thereby facilitating their investigations. In some such strains, designated high frequency of recombination (Hfr), the F plasmid was integrated into the bacterial chromosome. In others designated F prime (F'), various pieces of the chromosome were integrated into the F particle.

∼

F is not the only plasmid found in bacteria, nor is it the only one capable of transferring itself to another cell. Indeed, most bacteria contain plasmids, although only a minority of them are capable of self-transfer. Some bacteria contain several different plasmids. Plasmids that are capable of self-transfer are masters of genetic exchange. Some can spread rapidly through a population of bacteria, especially if the plasmid carries genes that benefit its bacterial host.

A dramatic example of the extent and impact of such horizontal, plasmid-mediated gene transfer occurred in Japan shortly after World War II ended. A series of investigations in which Tsutomu Watanabe played a leading role led to new understanding of the gene-transfer process. After the war, sanitary conditions in Japan were poor, and the country experienced an outbreak of bacillary dysentery, the shorthand term for bacteria-caused dysentery usually one caused by strains of the *E. coli*-like bacterium, *Shigella*. The explanation for this outbreak was presumed to be simple enough: the strains of *Shigella* that had become prevalent in Japan were resistant to sulfonamides, the antimicrobial drugs then available to treat the disease, so infections by such strains were impervious to conventional treatment. There's nothing terribly mysterious about the appearance of antibiotic-resistant bacterial strains. As was already known at the time, mutations occur spontaneously when bacteria (and other organisms as well) make rare mistakes as they replicate their DNA, and also when certain chemical or physical agents react with their DNA. Some such mutations confer resistance to antibiotics and other antimicrobial agents by one of several routes: they can confer on the cell that carries the mutation the ability to inactivate the agent chemically; they can alter the drug's cellular target such that it is no longer affected by the agent, or they can simply pump the antibiotic out of the cell faster than it comes in, keeping the ship afloat by constant bailing. In the presence of antimicrobial substances, these originally rare mutant strains carrying a resistance-conferring gene soon dominate a population, because in the presence of the antimicrobial agent, they continue to proliferate even as the parental, sensitive cells cannot grow or are killed. That's the simple but tragic basis for the fact that many of our precious and once-effective antibiotics are rapidly becoming useless.

The *Shigella* problem in Japan seemed to have been solved when new antibiotic drugs such as streptomycin, chloramphenicol, and tetracycline became available. But by 1957 a curious and confounding new problem arose. Strains of *Shigella* began to appear that were resistant to all three of these new drugs and to sulfonamide as well. Even more curious was the apparent linkages among these resistances: strains resistant to more than one drug were more frequently encountered than strains resistant to only one. Patients treated with a particular drug unsurprisingly shed in their stools strains that were resistant to that drug, but quite surprisingly and alarmingly they were resistant to several other antimicrobial drugs as well.

Then a revealing set of experiments showed that if multi-drug-resistant and completely sensitive strains were cultivated together in

DOUBTS AND COMPLICATIONS

the laboratory, the sensitive strain became multi-drug resistant. Such transfer of resistance occurred between strains of *Shigella* and even between strains of *E. coli* and *Shigella*. Watanabe called this an example of "infective heredity."

Of course, such transfer was highly reminiscent of the transfer of the F particle that Lederberg and Hayes had studied. Further studies showed that a self-transferring plasmid that Watanabe called the "resistance transfer factor" (RTF), later simply called an R factor, was responsible. It behaved quite like the F particle in *E. coli*. This R factor had incorporated within it genes that conferred resistance to antibiotics, often several of them, thereby conferring multi-drug resistance to the strain that carried it.

A coherent story began to develop. Self-transferring plasmids are common among microbes. Some of them have the ability to transfer themselves only to strains of the same species; some between closely related species; and some, anthropomorphically called promiscuous, can transfer themselves between distantly related species. While these plasmids are replicating within their microbial host, genetic recombination between the plasmid's and its host cell's genome sometimes occurs, rather like the formation of F′ strains. When it does, the plasmid becomes capable of transferring its host's genes to another cell. Plasmids that acquire genes that confer advantageous capabilities—for example, resistance to antibiotics if antibiotics are prevalent in the environment—are favored and become dominant. But any of the cell's genes can be transferred by this means of horizontal gene transfer, which we call conjugation. The term, although descriptive of the process in which two cells come in intimate contact to transfer genetic material, is a bit confusing because conjugation in eukaryotic organisms is the prerequisite for vertical inheritance. In prokaryotes, however, it is a means of horizontal gene transfer, completely independent of reproduction.

Viral Transfer of Genes

While these studies on the drug-resistance transfer plasmids were going on in Japan, Joshua Lederberg, now an assistant professor at the University of Wisconsin, Madison, continued his study of conjugation along with his wife Esther Lederberg, also a molecular biologist, and a graduate student, Norton Zinder. Together they examined how widely spread conjugation was among bacteria, conducting experiments similar to those that led to Lederberg's discovery of conjugation between strains of *E. coli*. They found that gene transfer also occurred among strains of the closely related bacterium *Salmonella typhimurium* (now awkwardly called by purists *Salmonella enteritidis* serovar typhimurium), and of course they assumed such transfer was the same or similar to the type of conjugation they had found in their studies on *E. coli*.

Almost as a matter of course, they repeated Bernard Davis's bundling board experiment to establish that cell contact was indeed necessary for transfer to occur. To their surprise, however, it was not. Transfer did occur in the U-tube, thereby establishing that physical contact between the two strains was not necessary. Some gene-carrying substance had passed through the ultrafine fritted glass disk at the bottom of the tube that separated the parental strains. The U-tube experiment had eliminated conjugation as the mechanism of this genetic exchange. Assuming, therefore, that they had encountered another example of transformation, they proceeded to isolate the material passing through the disk that carried genetic information. Again to their surprise, the gene-bearing agent they purified was not soluble DNA; it was a virus that attacks bacteria—a bacteriophage or phage.

The Lederbergs and Zinder called this third means of horizontal gene transfer among prokaryotes transduction. Over subsequent years the contributions of a number of investigators revealed how viruses can sometimes act as interconnecting genetic conduits among cells.

Transduction is a consequence of a viral infection gone somewhat awry. Like all viruses, those called phages that infect prokaryotes disassemble within their host cell and replicate their component molecular parts, viral nucleic acid and viral protein, separately. Then later in the infection, the component parts are assembled into new, intact viral particles (called virions). In the case of many prokaryotic viruses, assembly involves merely wrapping the viral nucleic acid within a viral protein coat. The viral nucleic acid of phages and other viruses can vary: single or doubled-stranded DNA, or single or double-stranded RNA; here we'll restrict ourselves to DNA-containing viruses. Unsurprisingly, mistakes can occur. Sometimes, rarely, a piece of the cell's own DNA rather than viral DNA becomes encased in viral protein and is released. This odd hybrid structure (called a transducing particle), owing to its viral protein coat, retains the ability to "infect" another cell. When it does, it inserts the prokaryotic DNA from the cell in which it was formed into the cell it infects, thus transferring DNA from one prokaryotic host (the first infected) into a new one (the subsequently infected). This mistake in viral replication results in horizontal gene transfer from one prokaryotic cell to another, from the cell in which the transducing particle was formed to the one it subsequently infected. The agents that passed though the Lederberg's bundling board were transducing particles, reminiscent of their original designation, filterable viruses.

This type of transduction is called generalized transduction because any of the cell's genes can be transferred. But if the virus genome becomes integrated into its host's chromosome, a different kind of transduction, called specialized transduction, can take place. When such a viral genome leaves the chromosome to become a virion, it sometimes brings adjacent chromosomal genes along with it to become a specialized transducing particle.

Unlike eukaryotes, prokaryotes have evolved multiple means (transformation, conjugation, and transduction) of transferring genes among neighbors. Although it is not yet fully understood, there must be selective advantages to this seeming genetic chaos. Such transfers are, however, somewhat controlled. Prokaryotes have also evolved mechanisms to detect and destroy DNA being offered by their neighbors.

CHAPTER 7

Can the Receiving Cell Say No?

Not all foreign DNA is benign. The recipient cell may be damaged during any type of horizontal gene transfer Some, notably viral DNA, can be lethal to the receiving cell, and certain plasmid DNA can be irreparably harmful. Randomly accepting DNA from the environment could jeopardize a cell's genetic integrity. Unsurprisingly, prokaryotes have evolved elaborate strategies for protecting themselves from genetic invasion by recognizing which DNA is not their own and developing a means to destroy it. This subtle discrimination is a demanding biochemical challenge. The receiving cell must be able to recognize DNA from any external source as foreign and destroy it without damaging its own DNA. The several mechanisms that have evolved to accomplish this demanding feat share the same basic schema: mark the cell's own DNA in a distinctive way and destroy all DNA that does not bear that distinctive mark.

Restriction and Modification

The essence of this schema, like the deciphering of the syntax of the genetic code, was discovered first by abstract genetics, and again simply by looking at plaques formed by phages in a lawn of bacterial cells on a Petri dish. The defining experiments took place in the early 1950s at Cal Tech in Pasadena, California by two recent arrivals from Europe, Giuseppe Bertani and Jean-Jacques Weigle. Weigle had left his position at the University of Geneva after suffering an early heart attack for the sunnier climes of California, where he began working with Bertani, a microbial geneticist from Italy. They were intrigued by an apparent genetic fickleness of bacterial viruses that they had termed "host-controlled variation," meaning that bacterial viruses (phages) exhibited characteristics that depended on which bacterial strain they had last infected. For example, almost every virion (individual virus particle) of the bacteriophage dubbed lambda in a population of viruses resulting from an infection of *E. coli* strain S formed a plaque (caused an infection) when plated with strain S, but only a minuscule number of them formed plaques on *E. coli* strain C. In contrast almost all lambda virions resulting from an infection of strain C formed plaques on strain C. In other words, *E. coli* strain C could detect a difference between lambda virions grown on S strains and those grown on C strains, lethally accepting those grown on a sister strain but rejecting those grown on an only slightly foreign strain. Recall that when a phage infects a bacterium, only its DNA enters its victim. Presumably, a particular strain of *E. coli* can determine that phage DNA produced in a different strain of *E. coli* is foreign DNA and destroy it, but phages grown on one of its fellow strain members are not detectable as foreign. Such DNA is not destroyed, and as a consequence that virion is lethal.

Bertani and Weigle's paper concluded that *E. coli* (and presumably other bacteria as well) are able to "restrict" (destroy) foreign DNA and to "modify" their own DNA (including phage DNA that they produce) in such a way that it becomes resistant to their own means of restriction. Many, perhaps all bacteria possess restriction/modification systems to exclude and thereby protect themselves against the entry of foreign DNA. As one would expect because each organism must distinctively mark its own DNA, there are a huge number of different restriction/modification systems among various prokaryotes.

Later others unraveled the biochemical basis of restriction/modification. Restriction is a consequence of enzymes called restriction endonucleases that recognize specific short sequences (four to eight base pairs long) and cut the DNA strand there, inactivating that sequence. Other resident DNA-destroying enzymes (nucleases) then attack the cut ends, destroying the entire molecule of foreign DNA. Modification is a consequence of enzymes that add methyl groups specifically to the site on the DNA where the restriction endonuclease normally cuts, rendering it resistant to such cutting. Each strain and species of bacteria has its own restriction / modification system for protection. By modification it can mark DNA as being its own. By restriction it can protect itself against invading foreign DNA.

Restriction endonucleases, hundreds of which have been purified from various bacteria and have been made commercially available, have played a crucially important role in the development of recombinant DNA technology because they cut DNA double stands at sequence-specific locations. The cut pieces can then be reassembled to produce recombinant DNA molecules, the fundamental underpinnings of recombinant DNA technology. These seemingly esoteric studies on how bacteria protect themselves from invasion by foreign DNA led to development of a major new industry.

Prokaryotic Immunity

Bacterial restriction / modification systems destroy whatever foreign DNA comes along, regardless of its source or history. They don't discriminate; all foreign DNA is recognized and destroyed. More recently, within this new century, an exquisitely precise mechanism, awkwardly called "clustered regularly interspaced short palindromic repeats" but referred to by its acronym, CRISPR, has been discovered that is quite different. It's extremely selective, acting as a primitive adaptive immune system. CRISPR destroys only DNA from a phage or a plasmid that a microbe bearing CRISPR has encountered and survived before.

Can the Receiving Cell Say No?

CRISPR constitutes yet another mechanism that prokaryotes have evolved to protect themselves from foreign DNA and, as a consequence, diminish the frequency of horizontal gene transfer.

The fascinating history of the discovery of CRISPR has been meticulously, although controversially, chronicled by Eric Lander of the Broad Institute of MIT and Harvard. The controversy surrounds the issue of who gets the major credit for the practical applications of CRISPR to gene editing, which we'll touch on later. The impact of these applications will be enormous. In 2014 *MIT Technology Review* called CRISPR "the biggest biotech discovery of the century." The stakes are high, and therefore the controversy is intense. Biologist Michael Eisen of the University of California, Berkeley called Lander's review "science propaganda at its most repellent."

The discovery of CRISPR is a tribute to the field of bioinformatics, more specifically to analyzing and comparing the DNA sequences of various organisms. The story begins in 1992 when Francisco Mojica, then a graduate student at the University of Alicante, Spain, noticed that *Haloferax mediterranei,* a halophilic archaeon he had been studying, contained a curious set of DNA sequences. He found multiple copies of thirty base-pair, nearly perfect palindromic sequences (reading the same in both directions, as in the phrase, "Able was I ere I saw Elba") that were separated by approximately thirty-six base-pair stretches. Over the next several years such sequences were found in other archaea and in various bacteria, including *E. coli.* Indeed, they are common. They're widespread among bacteria and almost universal among archaea. These are the sequences that came to be called CRISPR. But how can such sequences confer immunity?

The observation took on greater significance when such repeated sequences were found to be widespread among prokaryotes. They must serve some function, perhaps an important one. Why else would they be maintained in prokaryotic chromosomes, which unlike ours quickly discard unnecessary genes? Over the next few years a series of discoveries and speculations pointed the way toward understanding their function. Bits of phage or plasmid DNA were found interspersed between the repetitive sequences, thus suggesting a possible scenario: these bits might be residues of previous infections that the cells had survived, and the infectious DNA might be retained to protect in some way against subsequent infection by the same phage or intrusion of the same plasmid. These suppositions were validated when it was shown that altering the interspersed sequences decreased the resistance of the bacterium *Streptococcus thermophilus* to certain phages. *S. thermophilus* is used in cheese making, which is plagued by phage epidemics. The interspersed sequences of CRISPR did offer protection against specific phages.

Near the CRISPR sequence is a set of genes called CRISPR-associated sequences (*cas*). One of these genes, *cas9*, encodes a DNA-cutting enzyme, the kind called "nucleases" that are common in bacteria. But the Cas9 nuclease is quite unusual. Most DNA nucleases cut only one strand of DNA's double helix; Cas9 nuclease is bifunctional, cutting both strands. The rest of the CRISPR system guides Cas9 nuclease to a specific region of DNA, namely the DNA of an invading phage or plasmid. Because the CRISPR region incorporates bits of DNA from prior phage infections, RNA transcribed from it is complementary to the phage DNA from a subsequent infection. Owing to this complementarity, the transcribed RNA binds specifically to the incoming phage, guiding Cas9 with it. Cas9 cuts both strands of the phage or plasmid DNA so that the invader is destroyed, and the bacterium is protected. CRISPR offers yet another mechanism that bacteria have evolved to exclude foreign DNA and thereby suppress horizontal gene transfer.

Because of its remarkable RNA-guided specificity, the CRISPR-Cas9 system holds the promise of precisely editing genes to correct or replace faulty genes. It's only necessary to construct guide RNA that are complementary to the faulty gene. Then Cas9 will cut that gene, and if a piece of DNA encoding a properly functioning gene and overlapping the borders of the cut is present, it will be inserted at the site of the cut by the cell's own repair mechanisms. The faulty gene will have been replaced by a functioning gene. The CRISPR system holds the power of editing the DNA of any organism, including our own. All that's needed is the appropriate guide RNA, and large numbers of such guide RNAs are already commercially available.

The naturally occurring CRISPR systems have been modified to make them more useful for gene editing. Possibly the most important advance was made by Jennifer Doudna of the University of California, Berkeley in collaboration with Emmanuelle Charpentier, of the Hannover Medical School and Helmholtz Centre for Infection Research (HZI), Germany, and the Laboratory for Molecular Infection Medicine Sweden (MIMS), Umea University, Sweden. Naturally occurring CRISPR systems need two RNA molecules—one to recognize the foreign DNA and another to recognize Cas9. Doudna, Charpentier, and their collaborators linked these two RNA molecules together, vastly simplifying the system and rendering it much easier to apply.

Any human disease that has a precisely known genetic basis might well be amenable to treatment using the CRISPR-Cas9 system. Cystic fibrosis and sickle cell anemia, which cause lifelong suffering and early death, are dramatic examples. Even life-saving treatments, however, may be subject to serious ethical and practical concerns. Treatment would alter human germplasm, possibly leading to unanticipated catastrophes for future generations. We might inadvertently introduce new, damaging defects into the human gene pool. Do we have the right to treat and save the life of an individual at an even very slight risk of causing unknown damage to future lives? This is unknown territory.

The reality of such hazards is already with us. Were such treatment of humans to proceed in clinical practice, it would most probably be performed on embryos, in which the number of genes, including defective ones, are few. In May 2015, *Science* gave an account of experiments in China that used CRISPR technique to edit genes in non-viable human embryos. The results were sobering. Junjiu Huang and colleagues at Sun Yat-sen University in Guangzhou used the CRISPR-Cas9 system to edit the *HBB* gene, which encodes the human Beta globin protein in eighty-six human embryos. They hoped to find a method of preventing newborns from inheriting the blood disorder known as thalassemia. Two days after injection, seventy-one embryos survived, but only four of fifty-four tested carried the desired genetic change. Of these, only some of their cells had been corrected. Even more distressingly, these cells had large numbers of off-target mutations in genes other than the one encoding Beta globin. The method will certainly be improved, but the unknown possibility of off-target mutations will always lurk.

CRISPR also holds great potential as an experimental tool to produce in mice specific mutations known as knockout mutations, in which a specific gene function is eliminated. Such strains of mice are invaluable for a variety of experimental purposes.

In recognition of their achievements, Doudna and Charpentier were awarded a $3 million "Breakthrough Prize in Life Sciences" in 2015, a prize funded by Mark Zuckerberg and other internet entrepreneurs. No doubt we will be hearing much more about CRISPR and genetic editing of disease. A mechanism evolved by bacteria to protect them from hostile invasion will most certainly play a major role in human affairs. Moreover, there is reason to believe that in the future the CRISPR system will become more precise and therefore more safe. The major contributors most probably will be microbes themselves, owing to their enormous genetic diversity. It's been shown that there are hundreds of variations of the CRISPR system among prokaryotes. Some of these might be able to do a better and more precise job of gene editing than the original CRISPR-Cas9 system. In fact, one called CRISPR-Cpf1, which incorporates a different sort of nuclease, shows great promise. The future of gene editing at the moment looks very bright, certainly for experimental purposes and agriculture and possibly even for human therapy.

The awesome potential and hazards of CRISPR-mediated gene editing have captured the attention of scientists as well as the public. On December 3, 2015, an extraordinary meeting was convened in Washington, DC, to discuss use of such gene editing in humans. Attendees included the National Academy of Sciences, the National Academy of Medicine, the Chinese Academy of Sciences, and the Royal Society of London. They agreed that it would be "irresponsible to proceed" at present, but as knowledge advances the issue "should be revisited on a regular basis." Although the group has no regulatory power, its international composition and prestige will undoubtedly determine the future of this research in most if not all countries.

Horizontal Gene Transfer among Prokaryotes

So as we've seen, there are at least three ways that genes can be transferred horizontally among prokaryotes: transformation, transduction, or conjugation, but we've also seen that prokaryotes have evolved powerful mechanisms for recognizing and destroying the entrance of such foreign DNA. Despite these defenses, some genes are successfully transferred. How evolutionarily significant are such transfers? Are they massive enough to confound or even overwhelm vertical gene transfer as a guide to evolutionary relatedness? How common is horizontal gene transfer among prokaryotes? What is the evidence that it has occurred? In essence there are three kinds of evidence: finding a stretch of foreign DNA within a bacterial chromosome, finding an individual or groups of foreign genes in an organism's genome, or finding evidence of residual bits of a DNA-transferring vehicle—a phage or a plasmid—that could have mediated horizontal gene transfer. All these findings depend on knowing the sequence of a prokaryote's DNA, and now there's a huge database of such sequences, allowing assessments of the prevalence of horizontal gene transfer among prokaryotes.

The first type of evidence, finding a stretch of DNA in a prokaryote's genome that looks foreign, depends on its being different from the rest of the cell's DNA. Alien DNA is usually recognized by its having a G plus C fraction of its DNA that is markedly different from the background chromosomal DNA. In general, the G plus C content of an organism's DNA is uniform throughout its genome, and the G plus C variations among strains of the same species are quite slight. So finding a stretch of DNA in a cell's genome that differs from the background is pretty good evidence that it has come from some other not-too-closely-related cell by horizontal gene transfer.

At first glance, it might seem an odd coincidence that the entirety of an organism's DNA would have uniform G plus C content in spite of the large variations found among organisms. It's as though each organism or group of organisms has it own particular signature or makes an individual choice to express a certain preference for a particular G plus C composition. Indeed, that is the basis of the phenomenon, a choice called codon preference. The preference is expressed by how an organism selects within the redundancy of the genetic code. As we've seen, the genetic code is highly redundant. The codons UUU and UUC both encode the amino acid phenylalanine, for example. Using one or the other has no impact on the composition of the proteins being made, but it quite obviously does affect the G plus C content of a cell's DNA—the greater the cell's preference for the

Can the Receiving Cell Say No?

UUC codon, the greater the G plus C content of its DNA. The same argument can be made for all the code's redundancies. But what sets or drives these preferences?

The codons UUU and UUC are indeed redundant, but only genetically, not biochemically. Each codon needs its own special molecule to recognize it and its own special enzyme to do the recognizing (the pair of them are called tRNAs and aminoacyl tRNA synthetases, respectively). Prudently, cells husband the quantity of these recognizing molecules that they make. To manage this, they specialize with respect to which of these recognition molecules they make in the greatest amount, and as a consequence which codons they "prefer." As a result, preferred codons work better, so over evolutionary time they accumulate in the cell's genome. Mutations toward preferred codons are selectively favored; those in the opposite direction are not. This sets the G plus C ratio of all the cell's DNA. If a cell contains a stretch of DNA that differs from the background, one has to conclude that it must have come by horizontal gene transfer from some other cell (with a differing G plus C content and set of preferences), and selective pressures have not yet had sufficient time to alter it.

A difference in its G plus C content is not the only evidence that indicates a stretch of DNA has been transferred (relatively recently in evolutionary time) by horizontal gene transfer. Even more direct evidence is the presence of viral-like genes bordering a stretch of DNA, clear evidence that the stretch had been transferred by transduction, that it had been picked up by a bacterial virus and transferred from another cell. Similarly, the presence of genes flanked by sequences that are typical of plasmids is evidence that such genes might have been transferred horizontally by conjugation. Also genes that are "kinship unique"—those not found in closely related organisms—are favored candidates for having been transferred horizontally, not vertically. In all cases the sequence of a cell's DNA must be known in order to search for evidence of horizontal gene transfer. But now there's a plentiful reservoir of such information.

The genomes of many organisms, including our own, have been sequenced or are in the process of being sequenced; the Department of Energy keeps a tally in their Genomes Online Database. As of October 2015, 65,737 genomes have been sequenced. Many more have been partially sequenced or are in the process of being sequenced.

Using such evidence, studies on the frequency of horizontal gene transfers have yielded startling, and in the eyes of some taxonomists, alarming results that bring into question the validity of any Tree of Life for prokaryotes. For example, one such study by Jeffrey Lawrence and Howard Achman on *E. coli* concluded that 755 (17.6 percent) of its 4,288 genes were acquired through at least 234 separate events by horizontal gene transfer, a minimal estimate the authors suggested because transfers from organisms with similar G plus C ratios would go undetected. Also, regions of altered G plus C content over time tend to drift toward a cell's typical background level as a consequence of the same selective force that determined the background level, a process called amelioration. So only genes that have been transferred relatively recently can be detected as foreign on the basis of having a G plus C ratio that differs from the chromosomal background G plus C ratio. The actual incidence of horizontal gene transfer might well exceed the observed rate by a considerable factor.

On the basis of many studies over the last couple of decades, there can't be much doubt that horizontal gene transfer is an integral and important part of prokaryotic evolution. Curiously, genes associated with defining a new species or an important new function seem to be favored candidates for horizontal gene transfer or, more probably, they are more likely to be retained after transferal.

Genes or sets of genes called pathogenicity islands, which encode disease causation, are a fascinating example of the consequences of horizontal gene transfer. They illustrate the frequency of horizontal gene transfer, and they show how prokaryotic evolution can occur in sudden bursts, not just slow accretion of new capabilities through accumulation of many mutations. An assemblage of new genes can be acquired simultaneously, and they will be maintained if they offer a selective advantage.

E. coli is a particularly interesting example of a bacterial species that has been altered materially by horizontal gene transfer of pathogenic islands. Some such transfers have been quite recent. *E. coli* is a long-term although minor component of the human gut and feces, making up only about 0.1 percent of the total number of microorganisms found there, but it is readily detectable in all individuals. Indeed, *E. coli* seems to have evolved to live in the gut, metabolically able to thrive on materials shed from its mucus lining. *E. coli* is widespread, perhaps universal, the normal resident of the guts of animals as diverse as turtles, birds, and humans, and it's limited to that habitat. Finding *E. coli* elsewhere in the environment—in a lake or a stream—is *prima facie* evidence of recent contamination by fecal material, perhaps human. That's the basis for testing water safety in most countries, including the United States. If *E. coli* is present, the water has been contaminated with feces and must be considered dangerous to consume.

Most strains of *E. coli* are utterly innocuous. They do their hosts no harm and probably some good. Indeed, their well-known benign reputation, at least among researchers if not the general public, is a major reason that *E. coli* has played such a dominant role in microbiological research as well as biotechnology, the reason that it's probably the most thoroughly researched and understood organism on the planet. The biotechnology industry cultivates *E. coli* in vast quantities to produce an array of products, including insulin, human growth hormone, tissue plasminogen activator (to treat heart attacks), and rennin for cheese making. (Strict vegetarians reject cheese made with naturally occurring rennin, obtained from the fourth stomach of young ruminants, but they accept cheese made with rennin produced by strains of *E. coli* into which a bovine rennin-encoding gene has been cloned.) Now even complex antibodies are being made in *E. coli* for therapeutic use in humans, an astounding technological achievement.

In spite of its overall honorable reputation, a few strains of *E. coli*, certainly a tiny minority, are pathogenic, some dangerously so. There is now incontrovertible evidence that these strains became pathogenic rather recently in evolutionary time, perhaps some within our own lifetimes, as a result of receiving "pathogenic islands" (chunks of disease-encoding DNA) from disease-causing bacteria by horizontal gene transfer. The type of disease these various newly generated pathogenic strains of *E. coli* cause depends on the particular pathogenic island they received.

Certain of these pathogenic strains of *E. coli* infect the urinary tract. Such strains of uropathogenic *E. coli* (abbreviated UPEC) contain four distinct pathogenic islands, presumably acquired by at least four distinct horizontal transfers of genes, which confer a constellation of disease-producing powers. These include the formation of protein structures called pili (somewhat like the F pilus we looked at earlier) that extend out from the bacterial cells and have the property of being able to attach specifically to certain cells in the urinary tract. They may, for example, attach to surface cells of the bladder firmly enough to avoid being washed away by the flow of urine. Other pathogenic powers encoded by these islands include the formation of toxins that damage the host cells and a trigger that causes host cells to internalize bacterial cells. Together, these acquired genes converted an innocuous strain of *E. coli* into a dangerously infectious one that causes countless bouts of painful illness. In most cases, fortunately, it's an illness that can be treated with antibacterial drugs.

E. coli strains that contain various other consortia of pathogenic islands cause infant diarrhea (enteropathogenic *E. coli*, EPEC), tourist diarrhea (principally enterotoxigenic *E. coli*, ETEC), or diarrhea of rabbits, pigs, calves, lambs, and dogs (rabbit pathogenic *E. coli*), among others.

The presence of pathogenic islands is not restricted to *E. coli*. Other pathogenic bacteria as well owe their disease-causing capacity to such regions of foreign DNA in their chromosomes. Pathogenic islands dramatically illustrate the extraordinary impact of horizontal gene transfer on microbial evolution. By a single or a few horizontal transfer events, a harmless strain of *E. coli* can be transformed into a dangerous and sometimes deadly pathogen. It has found a new environment in which to compete.

Perhaps the most startling of these pathogenic products of horizontal gene transfer is a strain with the somewhat drab designation of *E. coli* O157:H7 that suddenly appeared in 1982. The alphanumeric designation, which can be applied to any strain of *E. coli*, is based on the properties of two highly strain-variable macromolecules on the cell's surface. Differences in them can be detected by their immune response, that is, by serology, the same method Griffith used to distinguish various strains of pneumococcus and discover transformation. One of these surface proteins, a molecule on the cell surface, is designated O for the German *ohne Hauch*, or "without film." It occurs in almost 181 forms, O1 to O181. The other surface protein, the protein of the cell's flagella, is designated H for the German *Hauch* or "film" and occurs in 53 distinguishable forms, H1 to H56 with a few skips. The German terms are based on historical observations of a thin surface film, literally translated as a breath or a mist, that appeared in *Proteus* bacterial cultures.

E. coli O157:H7 first caused trouble with two almost simultaneous outbreaks (illness clusters attributable to a single source) of intestinal disease, one in Oregon during February and March and the other in Michigan in May through June. Both outbreaks were characterized by severe abdominal pain and initial watery diarrhea, followed by grossly bloody diarrhea with little or no fever. The onset occurred four days after victims ate hamburger at the same fast food chain. The causative agent, isolated from stools, was identified as *E. coli* O157:H7, a strain then unknown except for its having been isolated from a case of hemorrhagic colitis in 1975.

In 1992 a larger outbreak of *E. coli* O157:H7 occurred in the state of Washington, and again the source was hamburgers at a fast-food chain. In total, 501 cases were reported that resulted in 151 hospitalizations and three deaths. Forty-five persons developed hemolytic uremic syndrome, a serious kidney illness resulting from the toxic substances released from red blood cells as they are destroyed as a result of bloody diarrhea. The illness, more common in children, can injure kidneys and cause death. This outbreak was controlled by removing 250,000 potentially contaminated hamburgers, an action which prevented an estimated 800 additional cases.

In 1994 *E. coli* O157:H7 infection became a reportable disease. This led to a greater understanding of the magnitude of the disease's impact, which is considerable. Annually the infection causes 73,000 illnesses in the United States. Since 1982 there have been more than 350 outbreaks and at least 40 deaths. Most of the reported illnesses (52 percent) resulted from consuming contaminated food, but some occurred through person-to-person transfer (14 percent) and from contaminated water (3 percent).

Inadequately cooked, contaminated meat, especially hamburger because it is mixed in large batches, thereby magnifying the impact of a single contaminated source, is the major food source of *E. coli* O157:H7. Contaminated fruits and vegetables, notably apple juice and spinach, have been responsible for several outbreaks. The natural reservoir of *E. coli* O157:H7 appears to be infected cattle. They carry the bacterium but are unaffected by it because they lack the sites on their intestinal wall to which the bacterium must attach in order to cause disease. Other animals, including pigs, also serve as symptomless reservoirs. One outbreak from contaminated spinach was traced to feral pigs, which dug under a protective fence to feed on an agricultural field of the vegetable. Another outbreak from contaminated apple juice was attributed to the apples having been harvested from a field in which contaminated cattle had grazed.

E. coli O157:H7 acquired its pathogenic potential by receiving via horizontal gene transfer a set of disease-producing factors. These include the ability to produce a toxin that is called Shiga-like, identical to the toxin produced by disease-producing strains of the bacterium *Shigella*. The toxin was most probably obtained from *Shigella* by horizontal gene transfer. Another disease-producing factor acquired by horizontal gene transfer is the ability to produce on *E. coli* O157:H7's cell surfaces a protein (pilin) that allows the bacterium to attach to receptors on our intestinal lining (the receptor cattle lack that confers their immunity). *E. coli* O157:H7's array of pathogenic islands is a bit more complex than that of other pathogenic strains of *E. coli*. It contains at least ten distinct islands.

The set of pathogenic strains of *E. coli* shows us how horizontal gene transfer enables a bacterium to make an evolutionary leap as a result of one or a few sudden acquisitions, enabling it to exploit a new habitat and way of life. There is some evidence that bacteria with such metabolic capacities to exploit available opportunities are more likely to be transferred horizontally. Pathogenic islands are such an example. They allow the recipient bacteria's lifestyle to change rapidly from innocuous free-living to invasive pathogenicity, a totally new ecological niche.

From these sorts of studies we have to conclude that horizontal gene transfer among bacteria must be common enough and sometimes recent enough that its consequences are readily apparent. The implications of such widespread horizontal gene transfer among prokaryotes appear to devastate the dream of understanding their phylogeny. The ubiquity of such transfer even raises the question of whether the Darwinian dictum of descent with modification can be applied to prokaryotes. If horizontal gene transfer is indeed common and widespread, there's no such thing as linear descent among the various bacterial groups. Rather than having a treelike family history, prokaryotes' lineage would be better described as a mesh of interconnections, a complex anastomosis. The details would be almost impossible to trace and unravel, at least with the tools available to us now.

Rampant horizontal gene transfer places in doubt the significance of species as applied to prokaryotes. Frank Harold, a microbiologist and science writer, eloquently summed up the situation: "At the risk of some exaggeration, one can argue that all the genes in the prokaryotic pool make up a single common market, and organisms [species] represent nothing more than the samples that proved successful."

Yet despite all this gene flow, we can still recognize lineages when looking at certain genes. So although the species concept may not be attainable for the organism, the concept of core genome and accessory genes does provide a sense that the core set of genes is behaving like a species.

CHAPTER 8

Can the Tree Be Trusted?

Discovery of the prevalence of horizontal gene transfer among bacteria begs an even larger question. Is its resulting evolutionary conundrum restricted to bacteria, or does it significantly impact other groups of organisms? It's now clear that horizontal gene transfer is not restricted to bacteria; it occurs among other organisms and even across wide biological gaps, including transfers between kingdoms. Examples of such transfers between bacteria and archaea, between bacteria and yeasts (which are eukaryotes), and between bacteria and fungi are now well established. There are also examples of transfer of genes between bacteria and plants and transfer to animals, including ourselves. We probably shouldn't be surprised by this extraordinary genetic promiscuity. After all, DNA is DNA, regardless of its source. It's only necessary to introduce DNA into a different organism and avoid its defenses; the receiving organism has the capacity to incorporate the errant DNA into its own genome and possibly to express the genes it carries.

One bacterium, the tumor-forming *Agrobacterium tumefaciens* (now called *Rhizobium radiobacter*), which causes crown gall disease in plants, makes its living by routinely passing some of its genes to its plant host, thereby genetically commandeering it to do the bacterium's bidding and permanently altering the plant's genome. *A. tumefaciens* inhabits well-aerated soils, growing there unobserved as most other bacteria do. If, however, a nearby plant suffers a wound, the phenolic compounds released from the plant attract the bacterium, which migrates toward the wound. The plant stimulates *A. tumefaciens* to form a long pilus, much like the sex pilus born by F⁺ strains of *E. coli*. When the *A. tumefaciens* cell enters the plant's cut or abrasion, its pilus attaches to a plant cell, and then the bacterium passes a plasmid (designated Ti) with its set of commandeering genes directly into the genome of its plant host, and the invading genes become part of the plant's genome. These genes command the plant to become hospitable, to create a favorable environment for the invading bacterium, to house it and feed it. The commanding genes include some that encode plant growth hormones; these cause the plant host to proliferate, forming a gall, a favorable, protected space for the invading bacterium to multiply. Other genes on the Ti plasmid direct the plant to synthesize unusual amino acid, called opines, that the invading bacterium can use as nutrients. The plant that makes the opines according to the instructions of the inserted bacterial genes can't utilize them itself. They benefit only the bacterium.

Occasionally other bacterial genes by chance or by design become integrated in the Ti plasmid, and they too are transferred (horizontally) to the plant during infection, an example of cross-domain gene transfer. This system is utilized to engineer plants genetically. By standard methods of recombinant DNA technology, genes from any source can be inserted into the Ti plasmid, and *A. tumefaciens* will then transfer them into the plant's genome. A number of properties, including herbicide and disease resistance, have been inserted into plants in this way.

The *A. tumefaciens* example certifies that genes can be transferred horizontally across domains from bacteria to plants, but is such transfer frequent and widespread enough to confound the Tree of Life? It certainly must extend to the host range of *A. tumefaciens,* which infects many plants, including grapes, stone fruits, nut trees, sugar beets, horseradish, and rhubarb, but it represents a tiny segment of the plant kingdom. There's no reason to believe that horizontal transfer of genes from bacteria to plants is frequent enough to impact the Tree of Life.

We might ask the same question about animals. The answer is undoubtedly the same. There is, however, now solid evidence that two gene transfer events from bacteria to primitive eukaryotes and hence to both plants and animals did occur in the dim evolutionary past, a genetic gift that forever changed the course of evolution and the entire ecology of our planet. The evidence for this transfer, which we will consider in detail later when we talk about the origins of eukaryotes, is now overwhelmingly powerful. These transfers didn't occur by any of the three modes of bacterial horizontal gene transfer that we've just discussed. They most probably occurred by an uptake method that is the unique province of eukaryotic cells: phagocytosis ("eating cells")—the propensity to engulf and internalize foreign objects. Prokaryotes can't do this. They take from their environment only compounds that are in solution. Only eukaryotic cells can internalize or "eat" solid objects. Regardless of the mechanism, prokaryotic cells were taken up intact.

The two events in question occurred when presumed primitive eukaryotic cells consumed two kinds of prokaryotic cells: an aerobic (oxygen-utilizing) bacterium and a photosynthetic, oxygen-producing cyanobacterium. The capturing eukaryotic cells kept their prey bacterial cells whole, alive, and functioning within their cytoplasm, a phenomenon called endosymbiosis, rather like two metabolic slaves: one endowed the eukaryote with the capacity to do the work of aerobic respiration, and the other carried out the task of oxygen-yielding photosynthesis, the hallmark characteristic of green plants. You may recognize the names we use for the modern, highly altered descendants of these intracellular captive bacterial cells: mitochondria (respiration conferring), found in animals and plants, and chloroplasts or plastids (photosynthesis enabling), occurring only in plants.

Altered as they have become over the millennia of intimate cohabitation, these endosymbionts retain certain striking resemblances to free-living prokaryotic cells. They are dramatically smaller than their eukaryotic host cells; they are enclosed within two membranes, the unique cellular architecture of Gram-negative bacteria, and they contain DNA in the form of a circular molecule, as the chromosomes of the vast majority of bacteria do. Bacteria that lack the second membrane are termed Gram-positive. (Eukaryotic cells are surrounded by single membranes. Their DNA is organized in a set of linear molecules.)

This transfer of bacterial genes to animals as well as plants (presumably by whole-cell engulfment) conferred on the recipients the world-changing, bacterial legacy of Earth's present-day oxygen cycle. Oxygen was produced by photosynthesis and used for high-energy-yielding respiratory metabolism. The capacity to produce oxygen must have come before the ability to use it. Solid evidence shows that the Hadean, prebiotic Earth completely lacked oxygen. All the molecular oxygen (O_2) that now fills our atmosphere (21 percent of the total) has a biogenic origin, the consequence of those certain bacteria called cyanobacteria, which evolved to carry out oxygenic (oxygen-producing) photosynthesis. Their capacity was later augmented when they became integral to plants. Time markers for this ancient, more than three-billion-year-old metabolic advance are embedded in the fossil record and also apparent as a logical geological consequence of an early oxygenated atmosphere.

The fossil record is somewhat equivocal, as we might expect. Searching for bacterial fossils that are billions of years old is technically demanding, and only a few Hadean-eon rocks have survived to provide evidence. Most have been weathered away or returned to

Earth's magma by constant subduction of tectonic plates. The few such rocks that have survived are found in the Pilbara Supergroup, west of the Great Sandy Desert in the northwestern corner of Western Australia; in the Isua sequence of Southern Greenland, and in the Barberton Mountain land of Swaziland in southern Africa. Some of these rocks have been buried so deep that the fossils they might have contained would surely have been destroyed by the intense heat of Earth's interior, but some appear to have survived these many millennia under conditions that fossils could endure.

To search for ancient microbial fossils, suitable ancient rocks have to be sliced into thin sections and examined under a microscope, a far cry from picking up fossilized dinosaur bones. The results of this sort of ancient microbial fossil hunting have been astounding. In 1984 J. William Schopf of UCLA soundly shook the world of science when he reported finding fossils of Cyanobacteria in an early Archean chert in Western Australia. These rocks and their embedded microbial fossils are 3.5 billion years old, almost as old as the oldest known rocks (3.8 billion years old); later Schopf found many somewhat younger fossils.

Other scientists, however, have challenged the authenticity of such ancient fossils, alleging that they might merely be the consequences of bubbles trapped in the then-liquid chert. The rarity and singularity of these supposed fossils adds to the skepticism of some scientists. Microbial populations are uniformly characterized by abundance, masses of cells living in close proximity. Why then would these suspected fossils be so rare, and when seen, occur as isolated chains of cell-like structures? Why wouldn't there be masses of them? In spite of these concerns, to my eye, some of Schopf's images are so overwhelmingly like cyanobacteria that I'm convinced.

Regardless of the authenticity of the 3.5-billion-year-old fossils, the somewhat younger fossils that Schopf found in the Bitter Springs chert formation of the Amadeus Basin in central Australia cannot be denied. They look almost identical to micrographs of modern cyanobacteria. One micrograph of a string of four cells, two squared-off cells in the middle, cells with rounded ends at the termini, could pass for a micrograph of living cyanobacteria. No one has challenged the authenticity of these images in the chert formation that housed them for the last one billion years.

Modern cyanobacteria derive their metabolic energy through oxygen-producing photosynthesis. It's logical to presume that these ancient fossil cyanobacteria did the same. However, that's far from certain. Autotrophy, the using of carbon dioxide rather than organic compounds as a source of carbon, had to occur early in biological evolution. How else could adequate amounts of the organic compounds upon which most life depends have been generated? There are many ways other than oxygen-yielding photosynthesis for modern prokaryotes to get the energy they need in order to live on carbon dioxide. Chemoautotrophy, or deriving metabolic energy from an inorganic chemical reaction (for example, oxidation of reduced sulfur compounds such as hydrogen sulfide, H_2S), might have played a role in life's early evolution.

Moreover, oxygen-releasing photosynthesis is not the only way for light energy to be harnessed to enable autotrophic growth. Many other forms of photosynthesis can and still do exist. In essence, photosynthesis is a process of trapping light energy and using it to reduce carbon dioxide to make carbohydrates at the expense of some other reducing agent, for example H_2S, some other reduced sulfur compound, or even an organic compound. Oxygen-yielding photosynthesis, a biochemically more complicated process, probably developed later, perhaps when some organism, undoubtedly a cyanobacterium, evolved to be able to utilize always-abundant water (H_2O) as the reducing agent. The cyanobacterium employed the hydrogen component of water to reduce carbon dioxide (CO_2) and release, as a waste product, the oxygen component in the form of oxygen gas (O_2).

So the cyanobacterial fossils are a powerful indication, though not solid proof, of when biogenic production of oxygen began. Another sort of evidence derives from the unusual chemical properties of isotopes. Carbon, like many elements, occurs in more than one isotopic form. Isotopes are atoms that are chemically the same, occurring in the same position on the periodic table of elements (deriving from the root *isos,* meaning "equal," and the Greek *topos,* or "place"). They differ, however, in their atomic weight, containing the same number of identity-conferring protons and electrons, but varying in their number of weight-conferring neutrons. Carbon occurs naturally in three isotopic forms: ^{12}C, ^{13}C, and ^{14}C—carbon-12, carbon-13, and carbon-14. The superscripts indicate atomic weight. All three participate in the same chemical reactions, but the three isotopes are not identical. ^{14}C is radioactive and therefore not stable; it self-destructs. At a certain probability, a ^{14}C atom loses an electron along with a charge-balancing antineutrino, thereby becoming a stable nitrogen isotope.

The relatively recent discovery of ^{14}C has had a profound impact as a tracer of biochemical reactions and as a tool for dating archeological finds. In 1940 Samuel Ruben and Martin Kamen, both twenty-seven years old, were working in the radiation laboratory at the University of California, Berkeley, when they realized the analytical power of this radioactive form of carbon. I was fortunate enough to witness some of their early enthusiasm. In the summer of 1942, two years after the discovery, I was a chemistry student in a class taught by Samuel Ruben. It was a common practice then to punctuate lectures in beginning chemistry and physics classes with experiments performed by the lecturer's assistant at a laboratory bench at the front of the lecture hall. Ruben was then an instructor in the chemistry department, a non–tenure track rank junior to assistant professor. He didn't look much older than most of the class. He told us about ^{14}C, but he didn't mention his role in its discovery, and he said he would conduct an experiment to illustrate its power as a means of "tracing" chemical reactions. As props he had two flasks, each containing a clear, colorless solution, a Geiger counter to detect radioactivity, and a filter flask. He told us that one of the flasks contained ^{14}C in the form of a carbonate, and the other contained something else. Which was which? The Geiger counter clicked when pointed at one of the flasks but not the other. Then he mixed the contents of the two flasks, and a precipitate formed, which he collected in the filter flask. Was the ^{14}C carbonate in the precipitate, or did it remain in the liquid phase? Again the Geiger counter answered the question. It clicked vigorously when pointed at the precipitate on the filter but only sporadically when pointed at the liquid that passed through the filter. His simple experiment was a not-to-be-forgotten illustration of the power of a radioactive tracer of chemical reactions. One of Ruben's flasks likely contained sodium carbonate, which was radioactive, and the other contained barium chloride. When mixed together, barium carbonate precipitated out of the solution.

At the time he was teaching us chemistry, Ruben was deeply engaged in using this same approach to trace the initial reaction products formed by plants as they take up carbon dioxide during photosynthesis. He would make $^{14}CO_2$ available to the plant and then determine which plant constituents became radioactive. He was making spectacular progress and had already published two preliminary papers, but World War II intervened. Ruben went to work for the National Defense Research Council and became engaged in developing methods for measuring ambient concentrations of poisonous gasses, including phosgene ($COCl_2$), which had been used with devastating consequences during World War I and the Sino-Japanese War. On September 27, 1943, just about a year after his memorable lecture demonstration, a defective ampoule of phosgene with which Ruben was working imploded in his hand. Ruben inhaled a large dose of the gas, ran out of his laboratory, and collapsed on the lawn in front of the old red-brick chemistry building on the Berkeley campus of the University of California. He died the next day, one month after he had been promoted to the rank of assistant professor. Were it not for this tragic event, the cyclic series of chemical reactions through which carbon dioxide is fixed during photosynthesis might have been called the Ruben rather than the Calvin cycle, the Calvin-Benson cycle, or the Calvin-Benson-Bassham cycle, as it is now known.

Five years after Ruben's death, in 1948, Melvin Calvin and his coworkers, Andrew Benson and James Bassham, also in the chemistry department of the University of California, Berkeley, used the methods pioneered by Ruben to unravel the cyclical route by which almost all the carbon dioxide on the planet is fixed. (A few bacteria use one or another of five alternative pathways, but quantitatively they are minor in terms of the world's total carbon economy.) Each turn of the Calvin cycle supplies the organism with a carbon atom, which flows into metabolic pathways leading to all the cell's organic components and regenerates a compound (ribulose 1,5-bisphosphate) ready to accept another molecule of carbon dioxide via a reaction catalyzed by the most abundant enzyme on the planet, ribulose bisphosphate carboxylase (RuBisCO). Calvin and his colleagues published a description of the cycle, and Calvin was richly recognized for his achievements. In 1961 he was awarded the Nobel Prize in Chemistry. He was even pictured on a 2011 US postage stamp. His colleagues,

Benson and Bassham, criticized him for his lack of generosity in acknowledging their contributions. Bassham, who lived in a rooming house not far from mine, told me of his resentment. Benson stated that he had been fired by Calvin.

Ruben's colleague, Martin Kamen, who during his long life made many important contributions to our knowledge of the mechanism of photosynthesis, was generous in his praise of his collaborator and friend. "Ruben was responsible, almost single-handedly, for the growth of interest in tracer methodology which occurred at Berkeley in the years 1937–1938."

Carbon-14 also has contributed to archeological dating as well as tracing metabolic pathways. Because ^{14}C is constantly being formed in our atmosphere through bombardment of atmospheric nitrogen by cosmic rays and constantly decays by beta emission back to nitrogen, its concentration in the atmosphere does not vary much. However, when ^{14}C becomes incorporated into living material, replenishment ceases as decay continues. Knowing the rate of its decay (a half-life of 5,700 years) and correcting for known fluctuations in atmospheric content of ^{14}C, we can calculate the age of organic material at archeological sites by measuring the amount of residual ^{14}C it contains. Owing to its rather rapid rate of decay, however, ^{14}C dating is only useful for dating objects that are less than about 20,000 years old. It can't help us date the more ancient events, such as the beginnings of oxygen-yielding photosynthesis.

Carbon's other isotopes play an important role in dating biological events too. Both ^{12}C, its most abundant form, and ^{13}C, a rarer form, are stable. As a consequence, their usefulness for assessing ancient events has no time limits, making it possible to probe the origins of Earth's most ancient carbon-containing substances. Organisms that utilize carbon dioxide as their source carbon via any of the several CO_2-fixing pathways can, to a limited extent, distinguish between $^{12}CO_2$ and $^{13}CO_2$, exhibiting a slight but detectable preference for the lighter, more abundant form ($^{12}CO_2$) over the heavier, rarer one ($^{13}CO_2$). Thus a carboniferous deposit that contains more $^{12}CO_2$ than the amount present in the atmosphere can be assumed with some degree of certainty to have been the consequence of biological utilization of CO_2—most probably of photosynthesis. The first carbon-containing rocks that exhibit such enrichment are about 3.5 billion years old, near the same age of Schopf's oldest cyanobacterial fossils. Is this a remarkable confirmation, or perhaps coincidence? The data do fit: the fossils that appear to be cyanobacteria appear in rocks as old as those that show isotopic evidence of biological utilization of carbon dioxide.

There is competing evidence, however, in the form of a geological marker that indicates when Earth's atmosphere filled with oxygen. The marker of this event is the age of the deposits of oxidized iron (ferric oxide, appearing as red rust) we see worldwide on Earth today, with particularly large accumulations in such places as Western Australia and Minnesota. When oxygen entered Earth's atmosphere in sufficient concentrations, these deposits inevitably formed from Earth's then vast quantities of quite soluble reduced iron (the ferrous ion), but such deposits are only about 2.5 billion years old. Why is there a gap of a billion years between the time when presumable oxygen-producing cyanobacteria first appeared and sufficient oxygen accumulated in the atmosphere to form deposits of oxidized iron? Iron itself is undoubtedly the major part of the answer. This ubiquitous element occurs in huge quantities on Earth; in terms of mass it is the most abundant element comprising Earth's crust. Earth's core is almost solid iron.

Before oxygen-producing cyanobacteria evolved, Earth was an oxygen-free planet with its immense endowment of iron in the form of reduced (ferrous) iron, which occurs as water-soluble greenish salts. Ferrous iron served as a huge sump for oxygen. As cyanobacteria produced oxygen gas, it reacted immediately with Earth's reservoir of ferrous iron, forming the vast deposits of red iron oxide (ferric hydroxide) that we have today. Only when ferrous iron was depleted did oxygen begin to accumulate in the atmosphere, perhaps taking nearly a billion years to reach the approximate 21 percent level that exists today. There were other oxygen sumps as well. These included reservoirs of reduced sulfur and reduced nitrogen, which I'll explore later in more detail.

Initially, the accumulation of gaseous oxygen in Earth's atmosphere was an environmental disaster. Oxygen is highly reactive and therefore intrinsically quite toxic. Its appearance on Earth is called the Great Oxidation Event (GOE); some have called it the "oxygen holocaust" because many or even most of the anaerobic organisms, then the only kind present on Earth, were killed. As is often the case, there were critically important benefits to the disaster: an oxygen-containing atmosphere gave rise to an accumulation of atmospheric ozone, life's principal protection from the sun's lethal ultraviolet radiation. Of course, many exquisitely oxygen-sensitive microbes did survive and live happily today in Earth's many oxygen-free niches.

Oxygen's high reactivity provided a stunning evolutionary opportunity along with its hazards—the possibility of high energy–yielding metabolism. Extant organisms such as baker's yeast that are capable of both anaerobic and aerobic metabolism illustrate the huge energy advantages of oxygen-based metabolism. When growing in the absence of oxygen (by fermentation, thereby converting sugar into ethanol and carbon dioxide), yeasts derive a mere two molecules of ATP (the molecular currency of metabolic energy) from each molecule of glucose they utilize; when growing in the presence of oxygen (by aerobic respiration converting glucose into carbon dioxide and water), they derive over thirty molecules of ATP from each metabolized glucose molecule. For baker's yeast, oxygen-based respiration offers a fifteen-fold advantage. Aerobic metabolism offers organisms opportunities for myriad energy-demanding evolutionary advances. The energy-profligate human brain is a dramatic example. Our brain constitutes about two percent of our body's weight, but even when resting, it consumes 20 percent of the metabolic energy we expend. More prosaically, tallies on the energy flux of *E. coli* growing on glucose in the presence of air show that it generates more ATP than it needs for growth. The excess ATP has to be torched off in unproductive reactions called futile cycles.

Soon after oxygen began to accumulate, microbes appeared that had evolved to exploit this high-energy resource. Aerobic respiration provided new evolutionary opportunities, including the evolution of eukaryotes and eventually ourselves. At this point, bacteria had evolved to foster on their own a complete oxygen cycle. Certain bacteria, the cyanobacteria, became oxygen producers, and others became the oxygen respirers.

As we've noted, plants and animals acquired the abilities to carry out oxygenic photosynthesis and oxygen-dependent respiration in a very direct way: one of their evolutionary predecessors simply engulfed a bacterium that possessed these capacities, and its progeny have continued to keep the bacterium's progeny as intracellular metabolic slaves. The captured oxygen producers became the chloroplasts. Chloroplasts retain their capacity to mediate photosynthesis in plants. The oxygen utilizers became the mitochondria found in both plants and animals, indeed in almost all eukaryotes. Of, course by these processes of bacterial ingestion, all the captured bacterium's genes were transferred to their new plant or animal host, representing a considerable horizontal transfer of genes. Most of the bacterial genes did not remain within the intracellular resident, however. They migrated to the host's own genome.

An Evolutionary—and Intellectual—Leap

Acquiring mitochondria and chloroplasts by horizontal gene transfer constituted a massive evolutionary leap for plants and animals. Discovering and proving that it had happened constituted a comparable and intriguing intellectual leap for science. The discovery of the microbial origin of mitochondria and chloroplasts, like the discovery of plate tectonics, required ample imagination and talent: simple observation led to whimsical speculation, to more serious consideration, and eventually to scientific proof. Long before plate tectonics became established or even considered as anything less than fanciful, speculation was rampant. How could anyone look at a world map without noticing that the east-facing bulge of South America seems to fit so neatly under Africa's west-facing bulge? Reasonable as such speculation might appear, there was no scientific support for a separation of the land masses. Conventional wisdom decreed that continents don't move. Not until the early decades of the twentieth century did the concept of continental drift become respectable, and not until the late 1950s and early 1960s was its driving force, sea-floor spreading, discovered. Shortly thereafter the patchwork of moving plates was mapped and their previous locations deduced. The plates have been moving for the past three billion years at speeds up to 100 millimeters a year. South America was indeed once attached to Africa and torn apart some 600 million years ago when the supercontinent, Pangea, broke apart. Scientific proof eventually caught up with and confirmed fanciful speculation.

Jan Sapp, a historian of science, has traced a similar pathway leading from speculation to proof that mitochondria and chloroplasts are the modern products of bacteria captured long ago. The story begins in the nineteenth century when it was noted that mitochondria and chloroplasts viewed under the microscope looked rather like bacteria. In 1883 the German botanist Andreas Schimper became the first to publish such an assertion: mitochondria and chloroplasts (a word that he had coined) are the intracellular progeny of formerly free-living bacteria captured long ago. Later in the 1880s the microbial origin of chloroplasts gained plausibility with studies that showed that the green color of certain marine animals, such as sea anemones and hydras, was a consequence of their harboring intracellular green algae that could be readily cultured free of their animal hosts. If algae could live transiently inside animal cells and contribute to their nutrition, why couldn't bacteria do the same on a permanent basis? Demonstrating their capacity for independent growth would prove their origin. In Russia in the 1890s, Andrei Famitsyn and undoubtedly many others who did not publish their results tried unsuccessfully to culture chloroplasts free of their host cell. These negative results, however, did not disprove the theory. If chloroplasts did indeed begin as free-living cyanobacteria, most of which are easy to culture, they must have lost the ability of exist independently during their long symbiotic dependency.

In the twentieth century, speculations about symbiotic origins of eukaryotic organelles became more frequent. In 1905 Ernst Haeckel added his prestige to the argument, proclaiming that chloroplasts might in fact be cyanobacteria living symbiotically within plant cells. Speculations continued about the possible symbiotic origins of mitochondria and chloroplasts, but all these proposals were just that—speculations. In 1925 a leading cell biologist of the time, E. B. Wilson, summarized the prevailing attitudes: "To many, no doubt, such speculations may appear too fantastic for mention in polite society; nevertheless it is within the range of possibility that they may someday call for some serious consideration."

Then such ideas fell out of favor in biology. They seemed to defy the Darwinian, then called neo-Darwinian, dogma that evolution progressed by multiple short steps, not great leaps, and further that the nucleus, not the cytoplasm, played the dominant role in determining metabolism.

By the 1960s, however, a cascade of morphological and chemical observations shifted the balance in favor of mitochondria and chloroplasts being intracellular bacterial symbionts. These included:

- Both organelles were surrounded by two membranes, a highly unusual biological structure seen elsewhere in nature only surrounding Gram-negative bacteria.
- Both organelles contain ribosomes, suggesting the capacity, now or in the past, of independent existence. Moreover, the ribosomes in the organelles are typical, in size and structure of their subunits, of prokaryotes, not eukaryotes.
- Both organelles contain DNA, and this DNA is circular, the form usually found in prokaryotes but never in the nuclear DNA of eukaryotes. Apparently these organelles carried their own life instructions, as all free-living organisms do.
- When these organelles divide, they pinch in two as prokaryotes do, not by forming the cross wall or developing the elaborate chromosome-separating scenario of mitosis typical of the way eukaryotic cells divide.

Taken together, these properties of mitochondria and chloroplasts made them appear to be prokaryotes living inside eukaryotic cells. The theory gained credence.

There can be little doubt now that the theory's acceptance rests on the publication of a fifty-page paper in 1967 by a twenty-nine-year-old biologist, Lynn Sagan, who later changed her name to Lynn Margulis. After at least fifteen rejections, the paper, "On the Origin of Mitosing Cells," was finally published in the *Journal of Theoretical Biology*. In it she suggested symbiosis as a major driver of evolution, a process she termed "endosymbiosis." She proposed a detailed mechanism for the genesis of eukaryotic cells through symbiotic relationships between prokaryotic cells. As we've seen, these proposals had been made before, and Margulis gave full credit to earlier advocates of the theory. Margulis's proposals, however, were more explicit. She proposed that an amoeba-like prokaryote had merged with an oxygen-utilizing prokaryote, resulting in a prokaryote containing a mitochondrion, thus making way for almost all eukaryotes, which with few exceptions are oxygen utilizers. Some of these symbionts merged with a cyanobacterium, thereby acquiring a capacity for oxygen-yielding photosynthesis; these became algae, which subsequently evolved to become higher plants. Sagan proposed a third endosymbiotic merger with a member of the group of corkscrew-shaped prokaryotes called spirochetes. She suggested that this combination had evolved to generate the complex eukaryotic flagellum and the structure of the nucleus, a proposal that has not withstood the test of time. Margulis stoutly defended all three of her proposals in print, journals, and books as well as in public forums. Later, as we will see, the first two were confirmed when new methods became available.

Margulis was a powerful advocate for her theories, as I experienced when I edited one of her submissions to *Microbiological Reviews*. Well-recognized for her work during her lifetime, she was elected to the National Academy of Sciences and awarded the National Medal of Science in 1999 by President Bill Clinton. She was also an advocate of positions that ranged from fringe to incomprehensibly fanciful. She was a vocal supporter of James Lovelock's rather squishy Gaia hypothesis that Earth acts as a sort of self-regulating organism that resists challenges. When questioned about how Earth could react to one of such challenges she famously responded, "Gaia is a tough bitch." Margulis took even more extreme views on other topics: she believed that AIDS was a consequence not of a virus but of a bacterium, a spirochete. And she believed that the destruction of the Twin Towers in Manhattan on September 11, 2001, was a consequence of their purposefully having been mined with explosives. In spite of holding some unsupported and controversial views, Margulis's contributions to biology are substantial and long lasting. as John Maynard Smith eloquently summarized: "Every science needs Lynn Margulis. I think she is often wrong, but she's wrong is such fruitful ways. I'm sure she's mistaken about Gaia, too. But I must say, she was crashingly right once, but many of us thought she was wrong then, too."

It was only a few years after publication of her 1967 paper until substantial evidence of her not being wrong about the origin of eukaryotes began to accumulate. The first evidence came in 1975 from Carl Woese and his colleagues. Using their original method of examining the products of T_1 RNase-digested small subunit RNA, they investigated the chloroplasts of the well-studied unicellular flagellate *Euglena gracilis*. By this time they had a good basis for comparison: they had examined the small ribosomal subunits of more than thirty prokaryotes and eukaryotes. They found that the digested product of the chloroplasts from *E. gracilis* is quite similar to the digested product of prokaryotes, not eukaryotes. This powerful evidence suggests that the chloroplast is the descendant of a prokaryote captured long ago. They further found that the digested products of the chloroplasts from *Euglena* and the red alga *Porphyridium* are nearly identical, suggesting their common lineage from a captured prokaryote. Margulis's hypothesis was then powerfully supported, if not unequivocally proven.

In the early 1970s, one of Woese's associates, Linda Bonen, joined the laboratory of W. Ford Doolittle at Dalhousie University in Nova Scotia, bringing with her the skills she had acquired from in Woese's lab. She, Doolittle, and Michael Gray used these techniques to study the mitochondria from wheat and came to the same conclusion that Woese had reached for chloroplasts: they were prokaryotic. Gray's laboratory, using newly acquired methods, went on to determine the complete DNA sequence of wheat mitochondrial DNA. This sequence proved to be most similar to a group of bacteria that Woese had designated alpha-proteobacteria. Certainly at that point the prokaryotic origin of both mitochondria and chloroplasts had been unequivocally proven.

The origins of mitochondria and chloroplasts are intimately connected to the origin of eukaryotic cells. That certainty leads us to the next question: What was the first cell that accepted a prokaryotic progenitor? Put in other terms, how did the eukaryotic cell evolve? Did one prokaryote merge with another prokaryote, or was it some sort of simple proto-eukaryote, capable of engulfing, that captured a prokaryote? If such mitochondria-free proto-eukaryotes existed, it's only reasonable to assume or hope their progeny—those that did not capture a prokaryote—would still be among us today. Indeed, groups of extant eukaryotes that lack mitochondria do exist. These amitochrondrial eukaryotic groups, as they are called, include the diplomonads and microsporidia, two groups of primitive, disease-causing eukaryotes. The notorious diarrhea-causing flagellate, *Giardia lamblia*, the bane of hikers and travelers, is a representative diplomonad. In 1681 Antonie van Leeuwenhoek observed and described a distinctive microbe that was almost certainly *Giardia* in his stools when he was suffering from diarrhea. In spite of this Leeuwenhoek is not credited with grasping the germ theory of disease, although he was certainly close to it.

It's tempting to leap to the conclusion that *Giardia* and related primitive eukaryotes are the modern-day descendants of the cells that captured the bacteria destined to become mitochondria. But such might not be the case. Indeed there is powerful evidence to the contrary. *Giardia*, as well as the other amitochondrial eukaryotes, carry in their chromosomes genes that come from the bacteria that became mitochondria in other cells. The most plausible and widely accepted explanation for these observations is that *Giardia* and its ilk once contained mitochondria, and some of the mitochondrial genes were transferred to the host cell's genome. When *Giardia*'s antecedents reverted to an anaerobic lifestyle, the mitochondria were lost. Still, one has to explain or rationalize why such loss of mitochondria is limited to a group of primitive eukaryotes. And the distinct possibility exists that bacterial genes, not the complete bacterial cell, were transferred to these organisms. Regardless of the mechanism, capture of bacteria leading to mitochondria and chloroplasts constituted a major horizontal transfer of genes from prokaryotes to eukaryotes.

Mysteries among the Archaea

Quite recently candidates for the first recipients of mitochondria and chloroplasts have been discovered among the archaea. This group, a putative phylum of archaea, was discovered by Thijs Ettema and his colleagues in Uppsala, Sweden. They dubbed the group "Lokiarchaeota" based on the location of their discovery—the cold ocean floor, 3,283 meters (about 11,000 feet) below sea level, some 9.3 miles from a venting structure on the hydrothermally active Mid-Atlantic Ridge between Greenland and Norway known as Loki's Castle. Appropriately, Loki was a mythical shape-shifting Norse god. No representative of Lokiarchaeota has been cultivated or seen; their existence and characteristics are known only from their DNA, which has been extracted from mud on the ocean floor and sequenced. Though unseen themselves, the information derived from their DNA has had a staggering impact. The Lokiarchaeota constitute a deeply branching line (a clade) of archaea most closely related to another primitive group of archaea known collectively as the TACK superphylum (Thaumarchaeota, Aigarchaeota, Crenarchaeota, and Korarchaeota). The remarkable finding about the Lokiarchaeota is the large array of genes they carry that are strikingly similar to eukaryotic genes. As we've seen, that in itself is not particularly unusual. Many archaeal as well as bacterial genes are similar to genes in eukaryotes, but the genes found in the Lokiarchaeota encode functions associated with eukaryotes not previously seen in prokaryotes. These genes, and there are hundreds of them, encode functions associated with the formation and movements of intracellular membranes, including those mediating phagocytosis (the ability of a cell to engulf outside material and form an internal membrane around it). Intracellular membranes are one of the defining characteristics of a eukaryote; prokaryotes are defined by the lack of them. These eukaryote-like genes, although so far found only in the mud, not in an archaeal cell itself, undoubtedly belong to the Lokiarchaeota because in all cases they are flanked by archaeal genes.

These genes that encode intracellular membranes make the Lokiarchaeota prime candidates for being the recipients of the bacteria that became mitochondria and chloroplasts, for being the progenitors of eukaryotic cells. They have many of the genes necessary for building the elaborate intracellular membrane system characteristic of eukaryotes, and they have genes that appear to confer the capability of phagocytosis, thereby providing a means of capturing the bacteria destined to become intracellular organelles. Ettema and his colleagues, whose paper announced the discovery of the Lokiarchaeota, suggested that they might be the "missing link" between prokaryotes and eukaryotes. Certainly the Lokiarchaeota fulfill this proposed link's major stricture: they possess characteristics of both predecessor and progeny.

If, indeed, the Lokiarchaeota prove to be the immediate progenitors of the eukaryotes, we would have to change our concept of the Tree of Life's trunk. Rather than archaea and eukaryotes being sister groups, as the paralogous rooting of the tree suggests, the archaea and eukaryotes would have a mother-daughter relationship. At the moment, both possibilities remain viable.

Conundrums in Plants and Animals

As we've seen, capture of bacteria leading to mitochondria and chloroplasts constituted an evolutionary bonanza for cells. As a consequence of a single event, a cell gained the capacity for aerobic respiration or photosynthesis, conferring an enormous selective advantage. Are there other examples in the plant and animal world that led to lesser evolutionary leaps? There are a few.

A fascinating example is the existence of aphids and certain other sap-sucking insects such as white flies. These insects, like all animals, including humans, are unable to make nine amino acids, the so-called essential amino acids. Humans must acquire them from our diet. It isn't that easy for insects, which live on plant sap. Plant sap consists largely of sugar; it is almost completely devoid of amino acids. Aphids nevertheless prosper, as all gardeners know, on this amino acid-deficient diet. They are able to do so because their antecedents captured bacteria that make amino acids for them. These captive microbes belong to the genus *Buchnera,* which are related rather closely to *E. coli,* and are readily apparent within certain specialized cells of aphids and white flies by light microscopy. Termed endosymbionts, the microbes look quite like their bacterial predecessors, although they, like mitochondria and chloroplasts, have lost their capacity for independent growth. They have, however, retained their ancestors' susceptibility to antibiotics. If aphids are treated with appropriate antibiotics, they lose their endosymbionts and become unable to exist on plant sap unless it is supplemented with the nine essential amino acids. During their long intracellular history, these endosymbionts have evolved to become amino acid–producing factories. Paul Baumann of the University of California at Davis, who initiated and pursued studies on their skills at producing amino acids, showed that they contain seven copies of the genes encoding the biosynthesis of the essential amino acid tryptophan, and they have lost the control mechanisms that prevent its overproduction. Captured bacterial endosymbionts supply sap-sucking insects with the essential nutrients that plant sap lacks, thus opening for insects new ecological possibilities.

This major cross-domain transfer of genes from bacteria to insects is not restricted to aphids and *Buchnera*. The half-inch leafhopper called the glassy-winged sharpshooter *(Homalodisca vitripennis)* is a notorious example. It has captured two nutrient-supplying bacterial endosymbionts that share their responsibilities. One of them, *Baumannia cicadellinicola* (named after Paul Baumann, who pioneered the work on aphids), satisfies the sharpshooter's need for vitamins. Another bacterium, *Sulcia muelleri,* supplies its essential amino acids. The glassy-winged sharpshooter is widespread. It, along with its two bacterial endosymbionts, infests around seventy different woody plants, including grapes, citrus, almonds, stone fruits, and oleander. With its needlelike mouthparts, the sharpshooter is able to penetrate into the plant's xylem and draw out its fluid. Such feeding is enough to cause significant damage to the plant, but because the sharpshooter is often a vector for the plant-pathogenic bacterium *Xylella fastidiosa,* such feeding can be deadly. As its name might suggest, *Xylella* is adapted to grow within the plant's xylem. It proliferates, plugging the plant's xylem and depriving the distal part of the plant of water. *Xylella* grows as a sticky layer (a biofilm) on the mouthparts of the sharpshooter, making it almost certain that each feeding will cause an infection. When *Xylella* infects grapes, it causes Pierce's disease. It also causes phony peach disease, oleander leaf scorch, and citrus X disease, depending on its host.

DOUBTS AND COMPLICATIONS

Pierce's disease has been a serious disease of grapes in California for over a hundred years, sometimes decimating vineyards in the southern part of the state and requiring some replanting in the northern part. It was chronic but manageable. At that time, the disease was transmitted by the blue-green sharpshooter. In 1989 the disease became much more serious when the glassy-winged sharpshooter was accidentally introduced on nursery stock from the southern United States. The blue-green sharpshooter was restricted to riparian regions and tended to damage leaves preferentially. It was responsible for some of the beautiful colors of vineyards in the late fall. In contrast, the glassy-winged sharpshooter moves quickly through all vineyards and can infect the trunks of the vines. Infected vines usually die within two years.

Within the past couple of years *Xylella* infection of olives has become a serious problem in Italy, although the spread of the disease is thought to be via infected pruning shears, not insects. All of this devastation begins with the intracellular endosymbionts *Baumannia cicadellinicola* and *Sulcia muelleri*, which supply the glassy-winged sharpshooter with supplemental nutrients that allow it to live on the contents of the plant's xylem and thereby transmit deadly *Xylella* from one plant to another. Once an intracellular endosymbiont is established, it exchanges genes with its host. Undoubtedly, this constitutes a major route of horizontal gene transfer between bacteria and insects.

These examples of bacteria-insect symbioses are at the extreme of a group of relationships called obligate endosymbiosis. Each partner is vitally dependent on the other, much as mitochondria and chloroplasts serve and are dependent on the cells that have engulfed them. The endosymbiont is no longer capable of independent growth, and the host's lifestyle would not be possible without the endosymbiont. There are myriad pattern variations of lesser dependency, all of which could lead to horizontal gene transfer between domains. Some of the most intriguing are the interactions of species of the bacterial genus *Wolbachia* and a vast variety of invertebrate animals, including fruit flies, shrimp, spiders, and parasitic worms.

Wolbachia is an intracellular parasite related to those bacteria that cause deadly typhus and Rocky Mountain spotted fever. It cannot survive outside its host cell because it cannot generate its own metabolic energy. Instead it steals and lives off some of the ATP that its host cell generates.

Wolbachia prospers in insects by manipulating the reproduction patterns of its hosts. Its principal route of reproduction is by vertical transmission from one host generation to the next, but it can only be carried in a female's eggs, not in a male's sperm; the sperm are too small to house them. Because *Wolbachia*'s reproduction and survival as a species depends on its being transmitted by females, it evolved to become a feminist, favoring their dominance in a population of insects. For *Wolbachia*, infecting a male is a dead end. By a number of routes, *Wolbachia* is extraordinarily successful in tilting its host population toward femaleness. For example, in the case of the beautiful African butterfly, *Acraea encendana*, orange with black and white wings, infection by *Wolbachia* has tilted the adult population to being 90 percent female. *Wolbachia* not only skews the gender ratio of its host species, it guarantees that the vast majority of females are infected. If an uninfected female mates with an infected male, most of her eggs will die. But an infected female can produce viable eggs after mating with either an infected or an uninfected male. Thus the fraction of infected females maximizes.

Wolbachia is widespread among insects. It is estimated to infect well over a million species, and its strategy for gender bending varies among them. In some species of wasps *Wolbachia* causes the females to produce only females, and these do not need males in order to reproduce. In other species males are born but become feminized and produce only eggs. In still others the gender balance is shifted more violently: *Wolbachia* selectively kills some of the males in the population. As a consequence, females do not have to compete with males for food, and in fact the dead males serve as a source of nutrition.

Wolbachia also infects filarial nematodes, including *Onchocerca ochengi*, the worm that causes river blindness, the world's major form of infectious blindness. By still-unknown mechanisms, *Wolbachia* establishes itself to be essential for the worm's survival. Antibiotics such as doxycycline and rifampin have been shown to be effective in controlling river blindness, presumably by the indirect route of killing *Wolbachia* and thereby debilitating and eventually eliminating the causative worm. However, for practical reasons it's easier to control river blindness by administering drugs that kill the host worm itself rather than eliminating its required symbiont. Undoubtedly the endosymbionts that insects harbor must constitute a major route through which genes are transferred from bacteria to insects and perhaps the plants that they attack.

There must be many other undiscovered examples of prokaryotic endosymbionts that mediate advantageous functions for various eukaryotes, thereby transferring genes horizontally. The biotechnology industry, which is often accused of breaking nature's fundamental rules by transferring genes from one domain to another, is certainly not the innovator of such exchange. Nature has been doing it for a very long time. The charge might better be hubris than playing God. We're not all that original.

It's a mystery why more such exploitations of prokaryotic capabilities by cell capture have not occurred. An interesting example is nitrogen fixation. A number of prokaryotes, including bacteria and archaea, have evolved the ability to fix nitrogen and therefore prosper in nitrogen-deficient environments, of which there are many. Growth of plants in many terrestrial and aqueous environments is limited by the supply of fixed nitrogen, but no plants evolved the capacity for nitrogen fixation. Certainly capturing a bacterium and maintaining as an endosymbiont a prokaryote capable of fixing nitrogen would offer that plant a prodigious selective advantage over its nitrogen-limited neighbors. But no plants harboring nitrogen-fixing endosymbionts are known to exist. As we have seen, legumes—beans, peas, and alfalfa—have evolved intimate and complex associations with nitrogen-fixing bacteria. So have certain plants such as cycads and well as alder trees and bayberry. In some cases these associations are highly evolved, both on the part of the plant and the prokaryote, but they stop short of establishing an endocellular symbiotic association. As we've seen, the well-known association between legumes and the nitrogen-fixing bacteria *Rhizobium* (and bacteria belonging to related genera) illustrates the intricate involvement of both partners that stops just short of taking the prokaryote inside the eukaryotic cell.

What Threat Is Horizontal Gene Transfer to the Tree of Life?

As we've seen, horizontal gene transfer is probably not sufficiently prevalent among eukaryotes to constitute a significant impediment to our understanding of the Tree of Life. But it is extremely common among prokaryotes. Should we just give up and recognize that the eukaryotic branches of the tree are rooted in a mesh of interconnected prokaryotes as some, notably W. Ford Doolittle, have suggested? Probably not, at least in my opinion and probably those of most practicing microbiologists. The metabolic core, the protein-synthesizing machinery of prokaryotes, which of course includes ribosomes, is rarely transferred horizontally. The vast majority of prokaryotic taxonomists have recognized this fact and have based their classification on it. All the recognized phyla, genera, and species of prokaryotes are based on ribosomal relationships. The burgeoning field of studies on the human microbiome, which we'll look at later, is based on that assumption. It's as though there is a central set of protein-synthesizing genes that define a prokaryote's cell identity, which it maintains and passes vertically while exchanging horizontally the genes making up the rest of its genome rather freely. With this caveat, a coherent branching Tree of Life is as good a descriptor of prokaryotic relationships as it is of eukaryotic ones.

But these considerations raise serious questions about the base of the Tree of Life. Was there no direct, traceable linage at life's earliest stages, before the modern protein synthesizing system evolved and locked in a cellular pathway? Most biologists believe that such was the case. So how did life begin? What was life's common ancestor? The question cannot be answered in any detailed manner, but modern studies have set the limits of plausibility of how the Tree of Life sprouted, and many detailed scenarios have been put forward.

{ PART THREE }

UNDERSTANDING THE TREE OF LIFE

CHAPTER 9

The Tree's Ecological Fruit

Deciphering the Tree of Life and developing the methods that have made this understanding possible has changed profoundly how we think about the field of biology and has widened its scope. Perhaps the greatest impact has been on microbial ecology, the interactions of microbes among themselves and in their environment. The microbial environments that have been opened to investigation are as diverse as deep-sea hydrothermal vents and our own bodies. Our expanded understanding has rested on advances in techniques to glean information encoded in the sequence of monomers in certain macromolecules, Pauling's semantides, but particularly on small subunit ribosomal RNA. By these methods it has become possible to locate particular organisms on the Tree of Life. Moreover, knowing the similarities and differences among the small subunit RNAs of organisms has led to the development of a technique, fluorescence *in situ* hybridization, that makes it possible to identify individual microbial cells and classes of microbes in their natural environments.

These methods were particularly important to the study of prokaryotes because other means of identification, as we've seen, depend on culturing them in the laboratory and testing them one at a time in various ways. Even this laborious approach cannot be used in most cases. Many prokaryotes resist laboratory cultivation entirely, at least by the methods available to us today. Estimates vary, but most microbiologists concede with some collective embarrassment that only a small fraction of one percent of the prokaryotes in most natural environments can be cultivated in the laboratory by methods now available to us.

As a result, determining which prokaryotes are present and what they might be doing depends on analyzing which macromolecules are present. This means of surveying leads to its own set of complications: many of the presumed species encountered by these means are new and have never been studied by conventional methods. The rules of prokaryotic nomenclature require a microbe to be cultivated before it can be named. Therefore, most such studies have summarized their results by listing which groups of prokaryotes were encountered. The major official divisions of bacteria are phyla. There are thirty of them. But the divisions that are most commonly used are the eleven designated by Carl Woese in 1987: Proteobacteria, Gram-positive bacteria, Cyanobacteria, Spirochaetes, Green sulfur bacteria, Bacteroides / Flavobacteria, Planctomyces, Chlamydiae, Radio-resistant micrococci, Green nonsulfur bacteria, Thermotogae.

The archaea are usually distributed among three phyla: Euryarchaeota, Crenarchaeota, and Nanoarchaeota, with representatives that have been cultured. However, a dozen or so other phyla have been proposed to accommodate representatives known only by sequencing the DNA in a particular environment.

Along with these advances in how we understand organisms' position on the Tree of Life came spectacular technical progress. Most notably, sequencing DNA became dramatically faster and cheaper. According to the National Human Genome Research Institute, in 2001 it cost about $8,000 to sequence a million base pairs (a megabase) of DNA. For the next six years that cost paralleled the predictions of Moore's Law (that microchips double in speed every two years) until in 2007 sequencing cost about $700 per megabase. As more new technology became available, costs dropped precipitously until 2011 and have continued to decline gradually ever since. By 2015 the cost had fallen to less than eight cents per megabase.

There have also been major advances in the ability to process and make sense of the raw data collected by sequencing the DNA present in a particular environment. Publicly available programs now allow us to process this mass of data and identify classes of microbes as well as determine their relative abundance. For the most part, these identifications are made very differently from the methods Carl Woese pioneered. He estimated the differences in the sequences of small subunit RNA among organisms in order to determine their relatedness. New methods of identifying the composition of a population of microbes in a particular environment involve sequencing all the DNA that is present, searching by computer for the DNA that encodes small subunit RNA, and using the information to identify the microbes or the classes of microbes that are represented.

Microbiologists have long been challenged to characterize the nature and diversity of microbial communities. Antonie van Leeuwenhoek had only a primitive handheld microscope in the 1680s, which showed him that the microbiota of teeth and feces were very different, but he had no way of knowing how or to what extent. Today microbiologists work with a precision that would have been unimaginable for Leeuwenhoek and continue to discover totally new classes of microbes. A spectacular illustration of this new potential is the recent discovery of a totally new class of methanogens. The methanogens that Woese and Wolf studied—the ones that led to the discovery of the archaea—all belong to the phylum Euryarchaeota. Paul N. Evans and his colleagues at the Australian Centre for Ecogenomics in Queensland, Australia, sequenced the DNA from a coal bed and found a new class of methanogenic archaea that belong to a phylum from which no representatives have yet been cultured.

The Microbiome

The US National Institutes of Health recognized these new opportunities and imperatives in 2008 by launching the $115 million Human Microbiome Project to study the consortia of microbes inhabiting various environments of our bodies, including oral, skin, vaginal, gut, nasal, and lung and how these might change in response to or as a cause of changing health. Microbiome, meaning the totality of a particular genetic microbial population, is a new word, a buzzword perhaps, replacing stogy but adequate terms like microbiota. Microbiome is the latest of a series of neologisms deriving from the *-ome* fragment of genome, the complete set of an organism's genes. That led progressively to proteome (all the proteins an organism makes), then a little more awkwardly to metabolome (all the metabolites or small molecules an organism contains or can make), and inevitably to microbiome (the complete array of microbes in a particular environment). The term microbiome was introduced by Joshua Lederberg to mean all the genes in a microbial population, but it has evolved and is often used as a synonym of microbiota, which is how I use it in this chapter.

The human microbiota is huge, consisting of 10 to 100 trillion diverse microbial cells spread over all our body surfaces, though concentrated mainly in the gut. Roughly 40 percent by dry weight of human feces consists of microbial cells. Most of the microbes there are bacteria, but also represented in lesser numbers are archaea, fungi, protists, and viruses. Viruses sometimes outnumber bacteria, but bacteria outweigh them. In total some estimate that there are at least tenfold more of them than our body's own human cells. And, of course, the microbial cells making up the human microbiota are far more diverse genetically than our cells, so the microbiome, strictly speaking, is much larger than our genome. One study estimates that it is over 150 times larger: 3.3 million different microbial genes as compared with our own mere 22,000. Moreover, the human microbiome is much more diverse than the human genome. One human genome is 99.9 percent identical to any other human genome, but a human microbiome may be up to 90 percent different from any other human microbiome. To further complicate matters, any human microbiome may vary over time as a result of health, diet, medical treatment, and other variables including, mysteriously, identity. It's not surprising that the National Institutes of Health should be interested in supporting research on the human microbiome. Some believe it offers a greater potential in the emerging field of personalized medicine than the much more highly publicized and better established Human Genome Project, which was initiated in 1990 and produced a first draft of the sequence of the human genome in 2001.

Care and Feeding of Our Microbiome

There can be no doubt that the microbiota of our gut impacts our health. Perhaps the most dramatic example of this impact occurs as a side effect of antibiotic therapy. Some antibiotics act against only a minority of types of bacteria. Penicillin, for example, even before resistant strains appeared, acted against only a few Gram-positive bacteria. In contrast, other highly effective antibiotics, including tetracycline, streptomycin, and chloramphenicol, kill a broader spectrum of bacteria, including both Gram-positive and Gram-negative species. The consequences of administering such a broad-spectrum antibiotic are not restricted to the target microbe and target organ. Through the bloodstream, antibiotics spread to almost all organs and kill indiscriminately. This profoundly changes the composition of the microbiota of the gut, sometimes eliminating almost all the bacteria that normally inhabit it. A common consequence of such massive slaughter is antibiotic-associated diarrhea, a benign, miserable, but self-limiting disease from which most patients recover spontaneously in a matter of weeks after the end of antibiotic therapy, presumably when the gut microbiota returns to a more normal composition. The side effects of are, however, a significant medical problem. An estimated 10 to 15 percent of hospital-treated patients develop antibiotic-associated diarrhea following broad-spectrum antibiotic therapy.

In certain cases the microbiota-altering consequences of broad-spectrum antibiotic therapy can be more serious. The killing of a significant portion of the body's normal microbiota opens niches for other microbes to occupy, some of which might be pathogens or become pathogenic when their numbers rise dramatically as normal competitors are suppressed. The yeast *Candida albicans* is such an example. *C. albicans* is a normal, benign member of the microbiota of the mouth. Following antibiotic therapy (and a variety of other triggers as well), *C. albicans* can become a pathogen, causing an annoying and sometimes painful disease called thrush, which is characterized by white patches in the mouth.

A much more dangerous infection that all too frequently follows broad-spectrum antibiotic therapy is caused by the anaerobic bacterium *Clostridium difficile*. It can cause a severe diarrhea, which can develop into a life-threatening inflammation of the colon. The traditional therapy for such an infection is to administer yet another antibiotic, vancomycin, to which most strains of *C. difficile* are sensitive, but such treatments are only marginally effective. Only about a third of antibiotic-treated patients are cured of the infection, and an increasing number of hospital-acquired strains of *C. difficile* have become resistant to vancomycin, rendering it useless for these cases. *C. difficile* is widespread in the environment. It's found in the air, soil, water, and even (benignly) in the intestines of some humans. Broad-spectrum antibiotic therapy creates an opportunity for it to become a dangerous pathogen. *C. difficile* infects over a half a million people and kills 15,000 of them each year.

In 1958, before *C. difficile* had been discovered, Ben Eiseman, a gastroenterological surgeon at a Veteran's Administration hospital in Denver, had a remarkable microbiological insight, the essence of which he published as five case reports in the journal *Surgery*. Several of his patients had become deathly ill after he had administered a broad-spectrum antibiotic prior to surgery, which was the standard practice. He speculated that his patients were suffering from the loss of their normal intestinal microbiota, which had been destroyed by the antibiotic. Logically, he reasoned that replacing the normal microbiota might cure his patients. He collected stool samples from a nearby maternity ward, reasoning that the young mothers were probably healthy, and added them to his patients' colons as an enema. It worked. All five of the treated patients regained good health.

Over the years, others have pursued the therapy initiated by Ben Eiseman, which has come to be known as fecal microbiota transplantation or FMT. It can be administered as an enema, a stomach tube, or orally in a capsule. FMT is spectacularly successful in treating *C. difficile* infections triggered by antibiotic therapy. Patients near death recovered within hours when treated with a single FMT. The results of randomized trials published in the *New England Journal of Medicine* revealed that fewer than a third of the patients treated with vancomycin recovered as compared with 94 percent of those who received FMT, the vast majority having received only a single treatment.

Crohn's disease, an autoimmune disease that affects over 700,000 Americans, introduces more complex questions about the relationship between intestinal microbiota and human health. This disease causes inflammation of the intestinal tract, which results in periodic intense abdominal pain and bloody diarrhea. It is conventionally treated with immunosuppressant drugs, many of which cause serious side effects.

When these do not work, it is sometimes necessary to remove part of the colon. Crohn's disease is considered to be incurable.

In certain respects the consequences of Crohn's disease resemble postantibiotic syndrome. The microbiota of patients suffering from the disease differs markedly from a normal microbiota. Returning it to a near normal state by FMT relieves the symptoms of the disease but not the disease itself. Over time the gut microbiota returns to its abnormal state and the symptoms recur. FMT must be repeated periodically, but from the limited amount of data now available, subsequent treatments seem to remain effective. Administering such treatments is complicated by the fact that the Food and Drug Administration has ruled FMT to be a drug, thereby requiring extensive testing, presumably for each source of the transfer. It's not clear why Crohn's disease distorts the colon's microbiota, but because FMT suppresses its symptoms it seems clear that the altered microbiota causes the symptoms of the disease.

∼

Although diseases like Crohn's have not yet yielded to the rapid advances in our understanding of the microbiota, a flood of information about how it develops in our bodies and how it impacts our health has become available through computer analysis of random DNA sequences in the microbiome.

Acquisition of the human microbiome begins early, and it changes rapidly. Of course, in the uterus a healthy fetus is microbe free, but as soon as twenty minutes after delivery a baby has developed a characteristic microbiota. Infants delivered vaginally have a microbiota characteristic of the vagina; those delivered by Cesarean section have a microbiota characteristic of adult skin. How the gut microbiota develops also depends on whether the baby is breast- or formula fed; it varies as feeding changes, reaching a composition typical of adults by about one year of age.

As we've noted, the human microbiome varies among individuals, who each bear a distinguishing microbiota. In a dramatic illustration of this fact, Rob Knight and colleagues at the University of Colorado showed in an arresting paper published in 2010 that a person's skin microbiota is sufficiently distinctive to be of forensic value, much like DNA evidence or fingerprints. It could even be said that this new type of evidence is a fingerprint of microbes. Knight and colleagues showed that the microbiota of our skin is almost as unique to an individual as a fingerprint. Only 13 percent of the species of bacteria on a person's hand are shared by any two people in the samples they examined. The singularity of our skin microbiota persists. Hand washing removes bacteria, but the same microbiota returns within hours.

Like a fingerprint, a sampling of our skin's microbiota is left on objects that we touch. Modern methods of obtaining and analyzing samples of microbiota have become amazingly sensitive. Knight and colleagues were able to characterize the microbiota left on a computer key by swabbing it, amplifying by PCR the approximately 1,400 small subunit-encoding ribosomal genes it contained, and sequencing these genes. The microbiota found on the computer key could then be compared with the microbiota of three individuals to determine which of them had struck the key. In a similar study, the individual who had used a computer mouse was identified from an database of 270 individuals.

Knight and his colleagues suggest that such characterization might become a powerful forensic tool that has some distinct advantages over the human DNA evidence so widely used today. DNA evidence can be obtained only if human tissue, blood, or semen is present on an object. Microbiota evidence requires only that the object has been touched. It didn't take long for this research to be reviewed in the popular press. *Time* noted in August 2015 that samples of skin microbiota are being evaluated in South Florida to investigate burglaries.

The distinctiveness and persistence of human skin microbiota that Knight and colleagues have documented, may explain the individuality of human odor. Dogs are particularly proficient in distinguishing people by odor. Dogs have been shown to be able to distinguish between identical twins by odor alone. Twins' genomes are identical, but their microbiotas must differ, although I have been unable to find any published research supporting this contention. There is, however, solid evidence that the skin microbiota effects and may even determine body odor. Some humans can also distinguish individuals by odor. I know individuals who can tell by odor which family member used a certain pillow. It is becoming increasingly clear that our microbiome identifies us. It also molds us.

The Gut Microbiome

The Tree of Life has told us a great deal about gut microbiota, which as we've noted varies among individuals. We've already talked about the dire consequences of an altered gut microbiota. It's becoming clear that microbes in our gut, our body's most important microbiota, play a much larger role in maintaining our daily body functions and our overall health. The gut microbiota contributes significantly to our ability to digest food. A number of substances that we are not able to digest and would otherwise pass through our gut unchanged become digestible because of our gut microbiota. Certain of the microbes in our gut attack nutrients that we can't, making some of their breakdown products available for us to utilize. The consequences of this fact are made abundantly clear by studies on mice.

"Germ-free" mice, those taken aseptically from their mother's uterus and raised in a microbe-free environment, become heavier after receiving a transplant of gut microbiota from a conventionally raised mouse. This phenomenon would undoubtedly also apply to humans who gain weight. A particularly clear example of specialization in gut microbes is the ability to digest porphyran, a carbohydrate present in certain red algae. Many Japanese who regularly consume red algae contain among their gut microbiota a bacterium, *Bacteroides plebeius,* that produces an enzyme that degrades porphyran into usable fragments, presumably enabling those whose microbiota contains that bacterium to derive energy from porphyran and thus from consuming red algae. The gut microbiota of Americans, who do not regularly consume red algae, lack *B. plebeius* and as a result the ability to derive energy from consuming porphyran.

There's a well-established relationship between composition of the gut microbiota and obesity, although it is not clear whether it's causal. The two major bacterial phyla in the gut microbiota are the Firmicutes and the Bacteroidetes. The Firmicutes are the collection of spore-forming Gram-positive bacteria and mycoplasma that we've met several times before; the Bacteroidetes are Gram-negative, non-spore-forming anaerobic bacteria that play important roles in breaking down a variety of nutrients. The two groups are the principal apparent variants, perhaps determinants, of the gut microbiota with respect to obesity. Several mutually consistent observations lead to these conclusions. First, genetically obese mice have a higher ratio of Firmicutes to Bacteroidetes in their gut microbiota than do lean mice. Second, it's well established that at least one of the causes of obesity in these mice is their better ability to extract nutrients from their diet. Finally, the feces of genetically obese mice, with their higher proportion of Firmicutes in their gut, contain less residual nutrients than the feces of lean mice. Composition of the gut microbiota does appear to have a causative relationship with obesity of mice. This supposition is strengthened by the observation that transferring the gut microbiota from an obese mouse to a germ-free mouse causes a greater weight gain than transfer of the gut microbiota from a lean mouse.

Another strain of genetically obese mice (TLR5 knockout) also seem to owe their obesity to the composition of their gut microbiota, but via a different route. Analysis of their feces shows no greater capacity to process the nutrients in the food; they simply eat more. These mice sometimes develop an ulcerative colitis resembling that of Crohn's disease. Their obesity can be cured by restricting their supply of food or by treating them with certain antibiotics that presumably alter the gut microbiota.

There seems to be little doubt that gut microbiota and obesity are connected in humans as well as in mice. If humans lose weight by restricting their intake of either fats or carbohydrates, the proportion of Bacteroidetes in their gut increases. So it seems that in both humans and mice a high ratio of Bacteroidetes to Firmicutes is associated with leanness and the reverse ratio is associated with obesity. Cause and effect are still difficult to assess. Does a high Bacteroidetes to Firmicutes ratio cause obesity, or does obesity cause the high ratio?

There is indirect evidence from a different set of observations that a high ratio is at least a contributor to obesity. Obesity is heavily correlated with type 2 diabetes, which results from insensitivity to insulin, which in turn can be measured by the rate of disappearance of glucose, a reaction that is low in persons with diabetes. In 2012 a group of Dutch scientists studied the effect of the composition of gut microbiota on a person's capacity to lower blood glucose as measured by the rate of disappearance of glucose. They studied a group of obese men with lesser ability to lower their levels of blood sugar. Following an intestinal lavage, some of the subjects were given a fecal transplant from lean adults; others were given a transplant of their own feces. Their diets were not changed. The results were quite dramatic. The median rate of disappearance of glucose for the experimental group increased from 26.2 to 45.3 micromoles per kilogram per minute. The authors suggest that such transplants might be the basis of an effective treatment regime for diabetes.

The gut microbiota is clearly a sensitive indicator of health as well as a cause or indicator of illness. We are just beginning to understand these complex interactions, both in our gut and elsewhere in our microbiome, and there are strong indications that there is more to come. For example, there are recent reports of correlations between composition of the vaginal microbiota and premature births. An extensive study conducted at Stanford University on pregnant women showed a strong correlation between vaginal microbiota and preterm birth. The authors suggest that analysis of vaginal microbiota is an effective way of predicting preterm birth, although they did not discuss possible therapeutic regimes. Undoubtedly other aspects of our microbiota will prove to be critical to our health.

Fluorescence *in Situ* Hybridization

With knowledge of the Tree of Life, along with the advent of rapid methods of sequencing DNA, it became possible simply by examining the DNA in a particular environment to enumerate the relative numbers of the microbes that inhabit it, which is the basis for the new field of metagenomics. A technique called fluorescence *in situ* hybridization, or FISH, takes this field a major step forward. It allows the investigator to identify individual cells in a microbiome and observe how they interact with other similar or quite different microbial cells. It takes metagenomics down to the single cell level. It can answer many questions. Do all the cells in a cluster, for example, belong to the same species? What is the relative abundance of the various species in a cluster? Are these associations uniform throughout the microbiome? The concept and technology of FISH are as simple as its potential is profound.

FISH depends on the ability of RNA molecules with similar or identical sequences to form stable double-stranded structures (to hybridize). FISH involves synthesizing a probe—a stretch of RNA about eighteen nucleobases long—that will hybridize with a ribosomal RNA of a specific species or group of microbes. A fluorescent molecule is then chemically attached to it. Next, the population of microbes to be examined is "fixed" (the microbiological euphemism for killed) by adding a chemical such as ethanol that renders microbes permeable to the probe. The probe is added, and after a period of time suitable for hybridization to have occurred, the sample is illuminated with ultraviolet light and examined. If hybridization has occurred with any of the cells in the sample, they will light up. Various microbes in a cluster can be made to light up in different colors by using different probes.

The particular power of FISH for microbial ecology depends on what Rudolf Amann at the Max Planck Institute for Marine Microbiology has called the "patchy conservation" of ribosomal RNA, referring to the tendency of certain regions of ribosomal RNA molecules to remain unchanged over considerable evolutionary distances. As a consequence, probes can be made that cause all bacteria, or all archaea, or all of a particular class of either one of them to light up. More specific probes for particular species of microbes can also be made. FISH can detect particular large and small groups of microbes that are present in a sampled environment, thus answering questions that conventional methods cannot.

Modifications have vastly increased the sensitivity of the procedure to the extent that tremendously subtle questions may be answered:

Which cells are dead? Which cells are actively growing? Which cells are able to utilize a particular nutrient? As an example of its power, FISH could assess the fraction of the microbiome of the Black Sea that is capable of the anammox (the major route for returning fixed nitrogen to the atmosphere) in spite of the fact that such microbes have defied attempts to culture them in a laboratory. The use and potential of this powerful method will undoubtedly continue to grow.

As the cost of sequencing falls and the ability to study microbes *in situ* increases, we are likely not only to fill in the twigs and branches of the Tree of Life, but to better understand how various microbiomes affect our health, perhaps our behavior, our immediate environment, and the health of the planet.

CHAPTER 10

The Tree's Beginnings

Any discussion of the Tree of Life inevitably leads to a tantalizing question: How did it all begin? We don't know, and perhaps we never will. We do know with some precision, however, when it happened—about 4 billion years ago. Comparatively speaking, life was generated extraordinarily rapidly. Earth was formed only about 4.6 billion years ago, and in its earliest days was undoubtedly an extremely hostile cradle in which to nurture life. Over the course of the twentieth century, scientists have proposed a number of scenarios for how life might have begun and have offered a plausible scientific rationale, if not absolute proof, for their validity. Studies on the origin of life have advanced from mere speculation to a respectable field of scientific inquiry.

Panspermia

Some students of the origin of life believe that the relatively brief period between 4.6 billion and 4 billion years ago was inadequate for life to have evolved, particularly during an era that was so environmentally hostile. This interval, known as the Hadean eon, was appropriately named for Hades, the Greek god of the underworld. It was hellishly tumultuous and hostile to life. Earth was being pummeled by meteors large enough to rip off the chunk of Earth that became our moon. Some of these doubters conclude that life must have begun someplace else on a more leisurely schedule and in a more life-friendly location. At some point, they believe, life was transported to Earth. This theory, called panspermia ("seeds everywhere"), posits that

The Tree's Beginnings

life is widespread in the universe, and that it came here from somewhere else.

Of course, panspermia raises some big questions it does not answer: How did it happen in that other place? How did it arrive here?

Some argue that there would be more time someplace else for life to evolve and also that life-generating conditions might be more favorable there. Francis Crick, the co-discoverer of the structure of DNA, made the interesting observation that life does bear certain marks of a foreign, unearthly origin. For example, molybdenum is a common cofactor for a number of essential enzymes for life as we know it. Chromium, with similar catalysis-enhancing potential, is not. Chromium, however, is ten times more abundant on Earth than molybdenum. Did life begin in a molybdenum-abundant location?

Even if life did originate elsewhere, getting here presents major problems. We know that objects like meteors strike Earth constantly, but it's hard to imagine that they are vehicles for any form of life. They are exposed to extremely high temperatures as they enter Earth's atmosphere. Most of them burn up. Smaller objects, such as a single bacterial cell, might not generate such intense heat, and they might move through outer space driven by solar winds, but they would be exposed to lethal doses of ultraviolet light during their long transit. Moreover, astronomers agree that it is highly unlikely that meteors from outside the Solar System have ever struck Earth. So panspermia would be limited to the Solar System, and there is no evidence for life's existing within it on any location other than Earth, certainly life's most congenial environment.

Despite these drawbacks, the history of advocacy for panspermia is so long and the list of its supporters so illustrious that the theory approaches the status of respectability. The Greek philosopher Anaxagoras subscribed to the idea of panspermia as did many extremely distinguished scientists until a little over a century ago. Its supporters included Hermann von Helmholtz, the nineteenth-century Prussian physician, physicist, and mathematician who clearly stated the law of conservation of energy, and the Irish physicist Lord Kelvin (William Thomson), who determined the value of absolute zero and for whom the absolute temperature scale is named. Kelvin famously declared, "we must regard it as probable in the highest degree that there are countless seed-bearing meteoric stones moving about through space. If at the present instant no life existed upon this earth, one such stone falling upon it might, by what we blindly call *natural* causes, lead to its becoming covered with vegetation."

Then more recently, in 1973, Francis Crick and the noted British chemist Leslie Orgel added their considerable prestige to the theory of panspermia. They conceded, as Carl Sagan vehemently argued, that life either as a passenger on a meteor or traveling alone could not be expected to survive passage through radiation-intense space and Earth's atmospheric barrier. They therefore presented a modified theory of "directed panspermia," suggesting that life had the help of intelligent beings to make its transgalactic voyage, that life came here on a protected space ship purposefully launched from elsewhere. They presented arguments that sufficient time has passed since the beginning of our universe for intelligent life to evolve elsewhere and for it to have become sophisticated enough to disseminate life abroad in the cosmos, including on planet Earth. There's no counter-argument to their proposal other than its being fanciful. (I've been surprised more than once by how distinguished scientists are attracted to the fanciful. I once heard the late Joshua Lederberg advocate spending resources to decipher meaning from the radio noise from Jupiter because that might be the only way the planet's inhabitants could break through the clouds of ammonia in its atmosphere to communicate with the universe.) Directed panspermia hasn't attracted many adherents. As we'll see, however, the arguments against panspermia don't apply to the building blocks of life, many of which have come from elsewhere in the Solar System, perhaps even the universe, via meteorites.

The Warm Little Pond

Panspermia has few supporters today. It seems more plausible that life began right here on planet Earth. Over a hundred years ago on February 1, 1871, in a letter to his friend Joseph Hooker, Darwin famously speculated about how life might have begun once, but never again. His theory has come to be known as "the warm little pond."

> *It is often said that all the conditions for the first production of a living organism are now present, which could ever have been present.—But if (& oh what a big if) we could conceive in some warm little pond with all sorts of ammonia & phosphoric salts,—light, heat, and &c present, that a protein compound was chemically formed, ready to undergo still more complex changes, at the present day such matter w^d be instantly devoured, or absorbed, which would not have been the case before living creatures were formed.*

The essence of the warm-little-pond proposal remains today the favored explanation of life's origin, if with many sophisticated variations and considerable experimental support. In a life-free environment, the constituents of living things accumulated through ordinary chemical reactions from the components of Earth's atmosphere at the time, and these assembled into self-reproducing organisms with built-in instruction for how to continue the process. The first part of the proposal is straightforward, uncontroversial, and can to a certain extent even be mimicked in the laboratory; there can be no doubt about it. It would have happened at least somewhere in a prebiotic world. The second part, the transition from the organic constituents, no matter how complicated, to the living organism, is a lot more difficult, and suggestions for how it might have happened are remarkably diverse and numerous. None is completely satisfying. How the spark of life was struck, if you like, has defied a single, definitive explanation.

Studies on the origin of life have attracted the interest of some truly remarkable and intriguing scientists. In the early twentieth century two quite different, almost exactly contemporary people, Alexander Oparin and J. B. S. Haldane, working independently, added chemical details and greater substance to Darwin's metaphor of the warm little pond.

Oparin was a chemistry professor at Moscow State University who put forward his ideas about how life might have begun in *The Origin of Life,* published in Russian in 1924. Oparin presented a logical sequence of events that could lead to life's beginning. Perhaps his most important contribution was the proposal that Earth's atmosphere, from which life's building blocks were formed, was in a reduced state. For example, carbon was in its reduced state as methane (CH_4), not in its oxidized state as carbon dioxide (CO_2). Oparin further proposed that the first organisms were heterotrophs living off the nutrients that had accumulated in life's primordial soup. Oparin's contribution was remarkably solid, melding information from astronomy, geology, chemistry, and planetary science, and it gained worldwide attention after the English translation was published in 1936. Antonio Lazcano, a present-day leader in the field, characterized Oparin's book as "a masterpiece in the area of evolutionary biology." In spite of his solid credentials as an exceptional scientist, Oparin went along with the Soviet system, even supporting the outlandish theories of Trofim Lysenko, who, with Joseph Stalin's support, waged a political campaign against genetics and its applications to agriculture, slandering and imprisoning those who spoke out against him, including his mentor, the geneticist Nikolai Ivanovich Vavilov. Oparin was a five-time recipient of the Order of Lenin.

J. B. S. Haldane was a gifted mathematician with many and diverse talents who made important contributions to our understanding of the physiology of hemoglobin and how it carries oxygen to tissues. Haldane, along with R. A. Fisher and Sewall Wright, was a founder of the field of population genetics, which has made important contributions to our understanding of how evolution works. Among his many insights Haldane predicted that the prevalence of sickle-cell anemia might be a consequence of its offering resistance to malaria, a proposal later confirmed by others. Haldane was a Marxist who equivocated about Lysenko but eventually rejected his theories. The Nobel Laureate Peter Medawar allegedly called Haldane the "cleverest man I ever knew." Haldane's ideas about the origin of life were formed entirely independently from those of Oparin. He published them before Oparin's book was translated into English.

Haldane's and Oparin's ideas about the origin of life are collectively referred to as the Oparin-Haldane hypothesis because of their similarities. The hypothesis is essentially an elaboration of Darwin's warm little pond, except the pond becomes the ocean and the reactants that give rise to Earth's primordial soup are identified as ammonia, hydrogen, and methane (according to Oparin), or carbon dioxide (according to Haldane). Haldane imagined that this prebiotic mixture was enclosed in an oily film, forming a cell-like structure. Oparin talked about "coacervates," or tiny colloidal droplets that would spontaneously form when certain organic compounds are dispersed in water, which served the same purpose as Haldane's oily film.

In 1953 the warm-little-pond hypothesis was famously put to an experimental test at the University of Chicago by a graduate student, Stanley L. Miller, working in the laboratory of Harold Urey.

Urey was quintessentially American. Born in Indiana, he was the son of a minister in the Church of the Brethren and educated in an Amish grade school. Urey was already a world-renowned scientist, having been awarded the Nobel Prize in Chemistry some nineteen years previously in 1934 for the discovery of heavy hydrogen, an isotope that Urey had named deuterium (D or ^{2}H, the only isotope—along with its next heavier form, tritium—that has a different name than the element's most abundant form). Urey had enriched deuterium sufficiently to make a spectroscopic identification by distilling naturally occurring hydrogen in liquid form, which, of course, contains all its isotopes. (I recall while doing undergraduate research under Robert Connick in 1947 in Gilman Hall at the University of California that there was a glass pipe perhaps a foot in diameter packed with the metal discs used to make shoe eyes which extended from the basement to the attic; it was the distilling column used to enrich for heavy water [D_2O or deuterium oxide]). Urey had also been a leader in developing methods to enrich uranium by centrifugation.

Based on his theories of the abundance of various elements on Earth, Urey concluded, as Oparin speculated, that its primitive, prebiotic atmosphere was reducing. This meant an atmosphere in which the major bioelements—nitrogen, carbon, and oxygen—were linked to hydrogen with some of the latter element left over as hydrogen gas. He concluded that the atmosphere was composed of hydrogen gas (H_2), ammonia (NH_3), methane (CH_4), and water (H_2O). This conviction led Miller and Urey to construct a sealed glass apparatus that they believed recreated the conditions of prebiotic Earth. In this they boiled a flask of water (mimicking the prebiotic sea); the steam rose through the gas mixture of the presumed prebiotic atmosphere and then entered a cold trap, which could be periodically sampled and analyzed chemically. An electric arc, mimicking lightning, was passed through the presumed Earth's prebiotic atmosphere.

In 1953 Miller published a brief, slightly over one page report describing the Miller-Urey experiment in *Science,* giving full credit to Urey for "many helpful suggestions and guidance in the course of this investigation." He describes the apparatus with the aid of a diagram that has been reprinted widely in textbooks and on websites dealing with the origin of life. Miller reported that the water in the flask became pink after the first day, then red and turbid by the end of a week, when the experiment ended. The experiment concluded with one noteworthy result: Miller was able to identify the presence of two amino acids, glycine and α-alanine, that are found in proteins along with another, β-alanine, that is not. He promised, "A more complete analysis of the amino acids and other products of the discharge are now being performed and will be reported in detail later."

In 1972 Miller and Urey repeated the experiment and found additional amino acids normally found in proteins, namely aspartic acid, glutamic acid, valine, proline, leucine, isoleucine, serine, and threonine, representing ten of the twenty found in proteins. Others reported that they too were able to synthesize a variety of amino acids under similar conditions, the stuff from which Darwin's "protein compound" could be formed.

All of the reports about nonbiological synthesis of amino acids agreed on one important point: the atmosphere must be reducing. If the reduced form of carbon, methane, is replaced by the oxidized form, carbon dioxide, amino acids are not synthesized. This becomes a serious sticking point because most scientists now believe that Earth's prebiotic atmosphere was not reducing. However, there are reducing microenvironments on Earth, even today. For example, the gases released from the hydrothermal vents at the mid-ocean ridges are in a reduced state.

If amino acids were not formed in quantity because Earth's prebiotic atmosphere was not reducing, panspermia of pond ingredients is another possibility for their origin. Amino acids must be present elsewhere in the Solar System because they are found on Earth in meteorites. In 1969 a huge meteorite struck near the town of Murchison, Victoria, in Australia. The Murchison meteorite shattered on impact, but fragments totaling in excess of 100 kilograms have been collected; one fragment weighs almost seven kilograms. These fragments contain the same amino acids as those that were formed in the Miller-Urey experiment and in approximately the same proportions.

Taken together, the results of the Miller-Urey experiment and its many replications along with analysis of meteorites leave little doubt that Earth's prebiotic oceans closely resembled Darwin's warm little pond.

Membranes

Of course, life does not exist as an oceanic solution. It always exists in the form of cells, and cell components are enclosed within membranes. Every cell on Earth arose by division of a pre-existing cell, and they are all bounded by membranes. The cell membrane divides along with the cell. Forming the first intact membrane with life's starting materials inside sounds like a major challenge, but such is not the case. There are compounds that spontaneously form spherical, balloon-like structures when suspended in water. These compounds, called amphipathic, are hydrophilic (water loving) on one end and hydrophobic (water hating) on the other. When dispersed in water, these molecules spontaneously arrange themselves into a double layer. Their hydrophobic ends seek the layer's water-free interior and their hydrophilic ends seek the water on both its inner and outer surfaces.

Fatty acids, the components of most fats and oils, are amphipathic. Their acid ends are hydrophilic, composed of carboxylic acid, or

The Tree's Beginnings

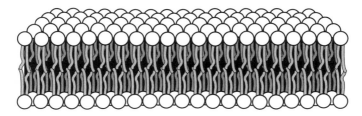

FIGURE 12. Sketch of how certain amphipathic compounds arrange themselves in an aqueous environment to form a double-layered membrane similar to those that surround most cells. Their hydrophobic tails, represented by the pair of wiggly lines, seek the membrane's water-free interior, and their hydrophilic heads (circles) seek the aqueous environment.

COOH, and their tails are composed of hydrocarbons (CH_3-CH_2-, and so on), which are hydrophobic. The membranes of most cells are made up of compounds called phospholipids, which are similarly amphipathic, forming two layers that create a water-free interior. David Deamer at the University of California at Santa Cruz has been a leader in the study of how life's first membranes might have formed. Early in his career he studied a phospholipid-forming enzyme. Watching the progress of the reaction, he noted, was quite startling. As phospholipid molecules were synthesized, they spontaneously coalesced into membranous structures, which when they became large enough, joined at their edges to form empty vesicles. It was as if cell-like structures were being formed as the product of this rather simple chemical reaction.

In the 1980s Deamer showed that meteorites contain lipid-like compounds that have the capacity to aggregate, forming stable membrane structures that resemble cells when suspended in water. Filling these structures with concentrated prebiotic soup also seems to occur almost automatically under conditions that might occur naturally. When they are dried, as we might imagine to have occurred on the edges of a small tide pool, for example, the spherical structures morph into layers, and of course the soup is concentrated. Then, when water is added back, as when water re-enters the tide pool, the layered structures rearrange into spheres with the concentrated soup inside them. These spheres are diverse in size and content; Deamer calls them protocells. One of these might have had life-forming potential. It's an intriguing thought, especially if we don't insist on the particulars. Deamer has said, "The first life was not just replicating molecules, it was an encapsulated system of molecules, a cell."

A cell membrane must do much more than just keep cellular contents together and concentrated. It must be semipermeable. That is, it must allow certain nutrients to enter and other materials to leave while keeping most of its contents from leaking out. To accomplish these disparate tasks, modern cell membranes are studded with proteins called permeases that render the barrier semipermeable. In general, a different protein is needed for each nutrient. That's a huge and complex task. An estimated 30 percent of *E. coli*'s proteins, for example, are needed to be inserted into its cell membrane to render it appropriately semipermeable. Deamer has experimented with incorporating short-chain fatty acids into membranes. He has shown that they confer a degree of semipermeability, nowhere near conferring the sophisticated complexity of a modern cell, but perhaps enough to get things started. Then, as almost all discussions of the origin of life conclude, natural selection must take over and lead the way—eventually—to optimized function.

The RNA World

At the heart of the confounding dilemmas facing those who study the origin of life lies Francis Crick's central dogma, which asserts that information stored in DNA flows through RNA to proteins that mediate the cell's metabolic activities, including synthesis of DNA. Genetic information flows in one direction and does not reverse from protein to DNA. This core principle for how all extant life functions contains a paradox, however: DNA is needed to encode the action of proteins, and proteins acting as catalysts (enzymes) are needed to make DNA. One can't lead to the other; the existence of each, it seems, depends on the other. One would conclude that both protein and DNA would have had to emerge simultaneously, which appears to be highly unlikely.

An alternative view is that RNA, not DNA or proteins, came first. In this scenario RNA would have played the roles of information storage and catalyst. RNA structurally resembles DNA, a long helical string of sugar and phosphate groups with protruding nucleobases. Of course there are fundamental differences. Instead of having the nucleobase thymine as DNA does, RNA contains the nucleobase uracil, which lacks thymine's methyl group ($-CH_3$), and instead of having deoxyribose as the sugar in its backbone, RNA has ribose. RNA usually occurs as a single strand rather than DNA's typical double strand, and RNA is less stable than DNA.

It is well established that RNA can store genetic information. It does so briefly in all cells (in the form of messenger RNA) as instructions are passed down the semantide pathway from DNA through RNA to protein. Moreover, many viral genomes are composed exclusively of RNA. These include the catastrophic human immunodeficiency virus (HIV), which causes AIDS, and influenza virus. Intriguingly, influenza virus's RNA genome is arranged in eight individual, chromosome-like segments. These viruses, of course, are not organisms, which are able to reproduce themselves; they direct other cells to do the reproducing. But the information for all the multiple directions these viruses harbor is encoded and stored in the form of RNA.

That brings up the question of RNA's catalytic and structural capabilities. Can it fulfill protein's cellular roles? Presciently, in 1967, on the basis of its complex structure, Carl Woese predicted that some RNA molecules might have catalytic capabilities. Francis Crick and Leslie Orgel made similar predictions.

Proof that RNA could act as an enzyme came fourteen years later in 1981, when Thomas Cech at the University of Colorado in Boulder showed that an RNA molecule could indeed catalyze cellular reactions. Studying the protozoan *Tetrahymena,* he found that a piece of gene (called an intron) was cut out of the gene by a reaction catalyzed by the messenger RNA it formed. The RNA message acted as a biocatalyst, an enzyme.

Then two years later, in 1983, Sidney Altman, working with Norman Pace, showed that the RNA component of the RNA-protein complex, ribonuclease P (RNase P), is responsible for its catalytic activity. Like the *Tetrahymena* message, it too is an example of an RNA enzyme. Such enzymes were dubbed ribozymes. In recognition of their paradigm-altering achievement, Cech and Altman shared the 1989 Nobel Prize in Chemistry. Now over ten different ribozymes have been discovered. Among these is the large RNA subunit of ribosomes, which catalyzes one of life's most crucial reactions, the linking together of amino acids (forming peptide bonds) to make proteins.

These discoveries, along with the known properties of RNA viruses, established that RNA alone can mediate life's two basic processes. Like DNA, it can store information and can be precisely replicated, fulfilling the requirements of heredity; like protein, it can catalyze vital cellular reactions, fulfilling the requirements of metabolism. It seems only logical to conclude that a membrane-enclosed RNA molecule supplied with a pool of necessary nutrients could constitute a primitive living cell, and that such a life form might have been the first living thing. Although the supporting knowledge of RNA's potential was much thinner at the time, Carl Woese made such a suggestion in 1967 in *The Genetic Code: The Molecular Basis for Genetic*

The Tree's Beginnings

Expression. In subsequent years a number of other scientists made similar observations.

In 1986, when most of the pieces of the puzzle had been assembled, Walter Gilbert, who had shared the Nobel Prize in Chemistry in 1980 for having devised a feasible method to sequence DNA, published a one-page paper in *Nature* titled "The RNA World." He formalized these ideas about how a primitive cell might have come into being. "The first stage of evolution proceeds," he states, "by RNA molecules performing the catalytic activities necessary to assemble themselves from a nucleotide soup."

Certainly the concept of an RNA world as an intermediate in the pathway leading to present-day life adds dramatically to the plausibility and complexity of Darwin's concept of the warm little pond. However, all theories on the origin of life or the last universal common ancestor (sometimes known as LUCA) share this commonality: each step forward leads to more questions. In the case of the RNA world, the question is this: Where did the RNA come from?

There is a plausible but partial answer. The building blocks of RNA are called ribonucleotides: a nucleobase, uracil (U), cytosine (C), adenine (A), or guanine (G) is bonded to the five-carbon sugar (ribose), which is in turn bonded to a phosphate group. An RNA-based beginning demands a source of ribonucleotides. In 2006 Matthew Powner, Beatrice Gerland, and John Sutherland showed that at least some ribonucleotides (U or C, which contain pyrimidine) can be made from materials existing on prebiotic Earth under conditions that existed there. It is reasonable to extrapolate and assume all four ribonucleotides were present in a form that could join together to make an array of RNA molecules, one of which might be capable of replicating itself.

But again there are problems. Joining together (polymerization) of ribonucleotides to make molecules of RNA under prebiotic conditions (that is, without a catalyst) is a slow process, and it's reversible. RNA would have been constantly breaking down into its component ribonucleotides. The only driving force for synthesis of RNA would have been the concentration of ribonucleotides, and that would undoubtedly be quite small. It seems more logical that there must have been a constantly available force of energy driving an RNA world. To find such an abundant source of energy and the ingredients to form life, we need to look beyond the warm little pond to a much larger, deeper, and warmer location, near the hydrothermal vents that spew hot water and a chemical soup at the bottom of the ocean.

Did Life Originate at the Bottom of the Ocean?

For a number of reasons, hydrothermal vents called white smokers are an attractive venue to serve as life's cradle. These vents exist near the mid-ocean ridges, created from magma oozing from the crack in the spreading tectonic plates. They are sometimes called the "lost city" because they generate tubes of calcium carbonate up to eighteen stories high. Unlike the black smokers, which occur at the ridge, these vents are about eighteen miles away. The effluent from these vents also differs from that of black smokers. It is alkaline and cooler, in the range of 40 to 90°C, hot but comfortably within life's tolerated limits. The first reason for these being the venue of life's beginnings, perhaps a trivial one, is that the oldest prokaryotic branches of the Tree of Life contain a preponderance of thermophilic or heat-loving organisms. The second is that these vents are reducing environments, abundantly spewing hydrogen and reduced forms of carbon and nitrogen, the type of environment that Miller and Urey found essential for forming the building blocks of life.

Nick Lane, John Allen, and William Martin make a powerful argument for an origin of life near the hydrothermal vents called black smokers, however. Their location on the mid-ocean ridge provides a lavish source of energy to drive life-generating reactions, an ingredient that all hypotheses on the origin of life recognize as essential. First of all, hydrogen and carbon dioxide, a pair of compounds known to react and yield metabolic energy, gush out of these vents. More importantly, black smokers generate a natural proton gradient, that is, an abrupt difference in concentration of protons or acidity. Water gushing from the vents is alkaline, and the surrounding ocean water is slightly acid. That might seem rather unrelated to metabolic energy, but it isn't, as was shown by Peter Mitchell in perhaps the greatest intellectual biochemical leap of the twentieth century.

In the mid-1960s, the rather eccentric Mitchell and his long-time colleague, Jennifer Moyle, renovated Glynn House in Cornwall as a laboratory and research institute. Working on a shoestring budget, Mitchell and Moyle successfully showed how a proton gradient could be converted to metabolic energy and vice versa. In fact, they demonstrated how all metabolic energy depends on proton or ion gradients. Mitchell called his hypothesis the "chemiosmotic theory." It has proven to explain the basis of all exchanges of metabolic energy, and in 1978 Mitchell was awarded the Nobel Prize in Chemistry "for his contribution to the understanding of biological energy transfer."

As we've discussed, ATP is the currency of the biological realm for metabolic energy. It or compounds related directly to it drive all metabolic reactions that need an energetic shove. Conversely, all forms of metabolic energy generation yield ATP (or a relative) as the end product. We've also dwelt on the fact that aerobic respiration is the most prolific generator of metabolic energy. Peter Mitchell proposed how they were linked, how aerobic respiration could generate ATP.

By the mid-twentieth century, biochemists had established unequivocally that respiration occurred through reactions in the cell membranes of prokaryotes and in the inner membranes of mitochondria, which are evolutionarily the same structures. But how did respiration occur? Some of the most illustrious biochemists of the day applied a traditional approach to this most basic metabolic issue. They disassembled the membrane in an attempt to isolate the component of it that mediated the generation of ATP by respiration, oxphos. (Oxphos, in the jargon of the day, is an abbreviation of "oxidative phosphorylation," or using oxidation to phosphorylate ADP to make ATP.) When the membrane was torn apart, however, oxphos ceased. Participants in the "oxphos wars," as their competition was known, strove to find increasingly gentler ways of breaking the membranes so as not to destroy its active oxphos kernel ingredient. No one was successful.

Then in 1961 Peter Mitchell thought about the issue in a new way: perhaps, he suggested, the active ingredient is the intact membrane itself. If it is ruptured in any way, no matter how gently, oxphos ceases. He had a good reason for why the membrane must remain intact: he suspected that the source of metabolic energy is a gradient of protons across the membrane. The difference in proton concentration across the membrane is potential energy; break the membrane and the gradient no longer exists. It's rather like water in an elevated container, which functions as a source of potential energy that can be tapped (by a water wheel, for example) as it flows back to ground level. But cell membranes are impervious to protons. How could a gradient of them across the membrane be created, and how could its return to a ground state generate ATP?

Mitchell had almost no experimental support for his next intellectual leap. He proposed that as aerobic respiration proceeds in a cell membrane of a prokaryote, protons (hydrogen ions) are pushed out across it, either out of the cell, in the case of prokaryotes, or out of the eukaryotic mitochondrion. This creates a gradient, a source of potential energy, a driving force for protons to come back into the cell. He hypothesized further that the protons can re-enter the cell only through particular proteins imbedded in the membrane. As they do so, they generate ATP. These special membrane proteins are the cell's metabolic water wheel.

Mitchell's explanation was met with considerable skepticism. But bit by bit, experiment by experiment, Mitchell's seeming fantastical proposal was proven to be true. Of all the supporting experiments, one set of them stands out in my mind as particularly notable and decisive. Researchers learned how to rupture bacteria, separate the cell membranes, and cause them to reform as hollow chambers, called membrane vesicles. It became possible to form membrane vesicles with specific compounds inside them. They found that if bacterial membrane vesicles were made with ADP and phosphate on the inside, and acid was added to the surrounding liquid, ATP was formed inside the vesicle. Thus, in a purified system, the proton gradient formed artificially by adding acid to the surrounding liquid generated ATP, a convincing proof that Mitchell's overall proposal must be true. Over the years details about the components became clear. The membrane proteins that extruded protons and the protein that generates ATP as protons re-enter the cell, called an ATPase, were identified. This ATPase is fundamental to life as we know it. It is present in early forms of all three branches of the Tree of Life and was used to locate its root.

Chemiosmosis is fundamental to cellular energy commerce. It can be used to transport nutrients across the membrane barrier and even to turn flagella, enabling a prokaryotic cell to swim. Organisms that are incapable of any form of respiration still need a proton gradient; they generate it by operating the membrane-bound ATPase in

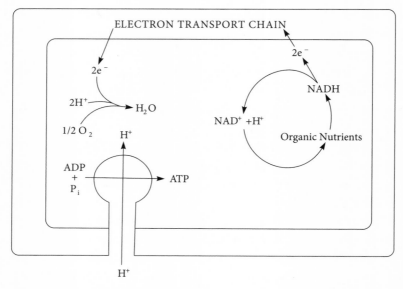

FIGURE 13. Peter Mitchell's chemiosmotic theory. Hydrogen atoms are removed from organic nutrients, becoming attached to a carrier compound NAD to form NADH, which donates a pair of electrons to an electron transport chain. As electrons flow through the chain, protons (H⁺) are extruded into the surrounding medium, creating a proton gradient across the membrane. At the end of the chain, the pair of electrons and protons react with oxygen-forming water. The proton gradient drives protons back into the cell through a membrane-bound ATPase, generating ATP from ADP and inorganic phosphate.

reverse—expending ATP they made by fermentation to force protons out of the cell.

The proposal by Lane, Allen, and Martin that life began near black smokers on the ocean floor depends primarily on the natural proton gradient that exists there, a potential for cellular chemiosmosis. They suggest that a membrane-bounded tube formed near the vent, and an alkaline effluent flowed through it, creating a proton gradient with the slightly acid seawater outside it. This provided an abundant source of energy to drive the formation of the RNA world. Then the tube budded into cells, and life was on its way.

Michael Russell, supervisor of the Planetary Chemistry and Astrobiology Group at the Jet Propulsion Laboratory in Pasadena, California, is a longtime proponent of the theory that life began at hydrothermal vents. He goes a bit further than Lane, Allen, and Martin, proposing that the soup of Darwin's warm little pond was made in these vents. Methane, carbon dioxide, and hydrogen react there to form acetic acid, the source of amino acids. He proposes that the first membranes might have been composed of amino acids.

Did Life Begin Twice?

Any theory about life's beginnings requires a certain amount of thinking outside the box, and few are as well-credentialed in that regard as Freeman Dyson at Princeton University's Institute for Advanced Studies. Dyson, perhaps the living definition of a polymath, wrote *Origins of Life* based on the Tarner Lectures he had given at Trinity College, Cambridge in 1985. His use of the plural "origins" in his title was intentional: he proposed that replication (as contrasted with reproduction) and metabolism are complicated but separable functions able to proceed independently, each more likely to develop on their own than simultaneously. A more probable scenario for the beginning of life, he suggested, would be for these two life functions to have begun separately in separate cells that later merged. The complexity of life's origins could be overcome, he submitted, once that life developed twice in simpler forms: one that was restricted to metabolism, and one that learned how to reproduce. The proposal attracted few adherents, myself included, but the clarity and logic of his presentation make the book well worth reading if not believing.

All in the Same Pond

In 2015 John Sutherland and his colleagues at the University of Cambridge took a major step toward a plausible explanation of how the incubating pond for life might have been simultaneously supplied with the necessary ingredients for its three essential component systems: replication, membrane compartmentalization, and metabolism. They showed how a single chemical system in the prebiotic world could yield much of the starting materials for the three systems. These include two of the nucleotides needed to make RNA, a precursor of membranes, and twelve of the twenty amino acids used in proteins. They were able to make all of them from hydrogen cyanide (HCN), a poisonous gas that they suspect might have spawned life.

Of course, that leads us to ask where the required hydrogen cyanide came from. Jupiter and Saturn—the planets, not the gods—are possibilities. These giant planets once orbited much closer to the Sun. As they migrated to their present location, they shed asteroids that pummeled young Earth some 3.8 billion years ago during the Late Heavy Bombardment. These intense collisions are presumed to have generated sufficient temperatures and pressures to have converted nitrogen from Earth's atmosphere and carbon from the meteors into the hydrogen cyanide that started it all.

Sutherland's remarkable contribution is in essence a sophisticated elaboration of Darwin's warm little pond.

"In the Beginning" Remains Just There

In the almost 150 years since Darwin suggested that Eden might lie hidden in his imagined warm little pond, we might with some considerable logic charge that science has made precious little progress toward determining the details of life's origin. There is not even an agreed-upon scenario to be probed and tested. Even if there were one that had accrued overwhelming supporting logic and experimentation, most probably proof is not possible. As the physiologist Jacques Loeb noted, we cannot experiment with the past. Precisely how life on Earth unfolded is an intrinsically unanswerable question. Even if we were to create life in the laboratory, there would be no way to prove that was how it happened originally.

But enormous progress has been made regarding the plausibility of Darwin's warm little pond. There is now a wealth of information, touched on only superficially here, about how the ingredients of life could be made in a prebiotic world, how they might have arrived from other places, how they might have been linked together to form macromolecules, and what natural materials might have served as catalysts for these vital reactions. Although the Tree of Life does not help us understand how life started, it constitutes an almost complete record of what happened subsequently. We still don't know our origin, but we do know in considerable detail how it could have happened and when it happened. Undoubtedly future research will enrich this pool of information, and I wouldn't be too terribly surprised if something meeting the general definition of life is made in someone's laboratory. Once life has started, even in its most primitive form, the powerful force of natural selection guides its improvement into the magnificent forms we all now admire. Chief among these are our relatives the multitalented microbes, upon which life continues to depend.

Further Reading

1. THE TREE'S MICROBIAL BRANCHES

Birrien, J. L., X. Zeng, M. Jebar, M. A. Cambon-Bonavita, J. Quérellou, P. Oger, N. Bienvenu, X. Xian, and D. Prieur. "*Pyrococcus yayanosii* sp. nov., An Obligate Piezophilic, Hyperthremophilic Archeon Isolated from a Deep-Sea Hydrothermal Vent." *International Journal of Systematic and Evolutionary Microbiology* 61 (2011): 2827–2831.

Darwin, Charles. *The Origin of Species by Means of Natural Selection, or, The Preservation of Favored Races in the Struggle for Life*. New York: Modern Library, 1993.

Hager, Thomas. *The Alchemy of Air*. New York: Broadway Books, 2008.

Specter, Michael. "Seeds of Doubt." *New Yorker*, August 25, 2014.

"Sur la Fermentation des Nitrates." *Comptes Rendus de l'Acacdémie de Sciences* 95 (1882): 644–646.

Suttle, Curtis A. "Marine Viruses—Major Players in the Global Ecosystem." *Nature Reviews Microbiology* 5 (2007): 801–812.

Vietmeyer, Noel. *Our Daily Bread: The Essential Norman Borlaug*. Lorton, VA: Bracing Books, 2012.

2. RELATIONSHIPS AMONG ORGANISMS

Crick, F. H. C. "The Biological Replication of Macromolecules." Symposium of *The Society for Experimental Biology* 2 (1958): 138–163.

Elias, Mikael, Korina Goldin-Azulay, Eric Chabriere, Julia A. Vorholt, Tobias J. Erg, and Dan Tawfik. "The Molecular Basis of Phosphate Discrimination in Arsenic-Rich Environments." *Nature* 1038 (2012): 11517.

Fitch, Walter M. and Emanuel Margoliash. "Construction of Phylogenetic Trees: A Method Based on Mutation Distances from Cytochrome *c* Sequences Is of General Applicability." *Science* 155 (1967): 279–284.

Pace, Norman R., Jan Sapp, and Nigel Goldenfeld. "Phylogeny and Beyond: Scientific, Historical, and Conceptual Significance of the First Tree of Life." *Science* 109 (2012): 1011–1018.

Sapp, Jan. *The New Foundations of Evolution*. New York: Oxford University Press, 2009.

Sarich, V. M. and A. C. Wilson. "Rates of Albumin Evolution in Primates." *Proceedings of the National Academy of Sciences USA* 58 (1967): 142–148.

Sarich, Vincent M. and Allan C. Wilson. "Immunological Time Scale for Hominid Evolution." *Science* 158 (1967): 1200–1203.

Schachman, H. K., A. B. Pardee, and R. Y. Stanier. "Studies on the Macromolecular Organization of Microbial Cells." *Archives of Biochemistry and Biophysics* 38 (1952): 245–260.

Watson, James D. *The Double Helix: A Personal Account of the Discovery of the Structure of DNA*. New York: Atheneum, 1968.

Wole-Simon, Felisa, Jodi Switzer Blum, Thomas R. Kulp, Gwyneth W. Gordon, Shelley E. Hoeft, Jennifer Petti-Ridge, John F. Stoloz, Samuel M. Webb, Peter Weber, Paul C. W. Davies, Ariel D. Anba, and Ronald S. Oremland. "A Bacterium that Can Grow by Using Arsenic Instead of Phosphorus." *Science* 332 (2010): 1163–1166.

Zuckerkandl, Emile, and Linus Pauling. "Macromolecules Documents of Evolutionary History." *Journal of Theoretical Biology* 8 (1965): 357–366.

3. ENTER DNA

Balch, W. E. and R. S. Wolfe. "Specificity and Biological Distribution of Coenzyme M (2-mercaptoethanesulfonic acid)." *Journal of Bacteriology* 137 (1979): 256–263.

Bemfield, Merton R. and Marshall Nirenberg. "RNA Codewords and Protein Synthesis: The Nucleotide Sequences of Multiple Codewords for Phenylalanine, Serine, Leucine, and Proline." *Science* 147 (1965): 479–484.

Brownlee, G. G., F. Sanger, and B. G. Barrell. "Nucleotide Sequence of 5S-ribosomal RNA from *Escherichia coli*." *Nature* 215 (1967): 735–736.

Crick, F. H. C., F.R.S., Leslie Barnett, S. Brenner, and Dr. R. J. Watts-Tobin. "General Nature of the Genetic Code for Proteins." *Nature* 192 (1961): 1227–1232.

Crick, Francis. *What Mad Pursuit: A Personal View of Scientific Discovery*. Alfred P. Sloan Foundation Series. New York: Basic Books, 1988.

Doi, Roy H. and Richard T. Igarashi. "Conservation of Ribosomal and Messenger Ribonucleic Acid Cistrons in *Bacillus* Species." *Journal of Bacteriology* 90 (1965): 384–390.

Elson, D. and E. Chargaff. "On the Deoxyribonucleic Acid Content of Sea Urchin Gametes." *Experientia* 8 (1952): 143–145.

Fox, G. E. and C. R. Woese. "The Architecture of 5S rRNA and Its Relation to Function." *Journal of Molecular Evolution* 6 (1975): 61–76.

Ling, J. R. "Robert E. Hungate's *The Rumen and Its Microbes* after 25 Years." *Applied Microbiology* 13 (1991): 179–181.

Marmur, Julius, Edna Seaman, and James Levine. "Interspecific Transformation in *Bacillus*." *Journal of Bacteriology* 85 (1963): 461–467.

Stanier, Roger Y., John L. Ingraham, Mark L. Wheelis, and Page R. Painter. *The Microbial World*, 5th ed. Englewood Cliffs, NJ: Prentice-Hall, 1986.

Stent, Gunther S. *The Coming of the Golden Age: A View of the End of Progress.* Garden City, NY: Natural History Press, 1969.

Woese, Carl R. and George E. Fox. "Phylogenetic Structure of the Prokaryotic Domain: The Primary Kingdoms." *Proceedings of the National Academy of Sciences USA* 74 (1977): 5088–5090.

4. THE ROSETTA STONE

Fox, G. E., L. J. Magrum, W. E. Balch, R. S. Wolfe, and C. Woese. "Classification of Methanogenic Bacteria by 16S Ribosomal RNA Characterization." *Proceedings of the National Academy of Sciences USA* 74 (1977): 4537–4541.

Mayr, E. "Two Empires or Three." *Proceedings of the National Academy of Sciences USA* 95 (1998): 9720–9723.

Woese, Carl R. and George E. Fox. "Phylogenetic Structure of the Prokaryotic Domain: The Primary Kingdoms." *Proceedings of the National Academy of Sciences USA* 74 (1977): 5088–5090.

5. FROM THE TREE'S ROOTS TO ITS BRANCHES

Copeland, Herbert F. "The Kingdoms of Organisms." *Quarterly Review of Biology* 13 (1938): 383–420.

"Default Taxonomy: Ernst Mayr's View of the Microbial World." *Proceedings of the National Academy of Sciences USA* 95 (1998): 11043–11046.

Garrity, George M., ed. *Bergey's Manual of Systematic Bacteriology*, 2nd ed. New York: Springer-Verlag, 2001.

Gogarten, Johan Peter, Henrik Kibak, Peter Dittrich, Lincoln Taiz, Emma Jean Bowman, Barry J. Bowman, Morris F. Manolson, Ronald J. Poole, Takayasu Date, Tairo Oshima, Jin Konishi, Kimitoshi Denda, and Masasuke Yoshida. "Evolution of the Vacuolar H^+-ATPase: Implications for the Origin of Eukaryotes." *Proceedings of the National Academy of Sciences USA* 86 (1989): 6661–6665.

Hug, Laura A., Brett J. Baker, Karthik Anantharaman, Christopher T. Brown, Alexander J. Probst, Cindy J. Castelle, Cristina N. Butterfield, Alex W. Hernsdorf, Yuki Amano, Kotaro Ise, Yohey Suzuki, Natasha Dudek, David A. Relman, Karl M. Finstad, Ronald Amundson, Brian C. Thomas, and Jillian F. Banfield. "A New View of the Tree of Life." *Nature Microbiology* 1, no. 16048 (April 2016); doi:10.1038/nmicrobiol.2016.48.

Iwabe, Naoyuki, Kei-Ichi Kuma, Masami Hasegawa, Syozo Osawa, and Takashi Miyata. "Evolutionary Relationship of Archaebacteria, Eubacteria, and Eukaryotes Inferred from Phylogenetic Tree of Duplicated Genes." *Proceedings of the National Academy of Sciences USA* 86 (1989): 9355–9359.

Lwoff, A. "The Concept of a Virus." *Journal of General Microbiology* 17 (1957): 239–255.

Margulis, Lynn and Karlene V. Schwartz. *Five Kingdoms: An Illustrated Guide to the Phyla of Life on Earth*. New York: W. H. Freeman and Company, 1998.

Pace, N. R. "Problems with 'Procaryote.'" *Journal of Bacteriology* 191 (2009): 2008–2010.

Pace, Norman R. "Mapping the Tree of Life: Progress and Prospects." *Microbiology and Molecular Biology Reviews* 73 (2009): 565–576.

Richards, Robert J. *The Tragic Sense of Life: Ernst Haeckel and the Struggle over Evolutionary Thought*. Chicago: University of Chicago Press, 2008.

Schwartz, Robert M. and Margaret Dayhoff. "Origins of Prokaryotes, Eukaryotes, Mitochondria, and Chloroplasts." *Science* 199 (1978): 395–403.

Stanier, R. Y. and C. B. van Niel. "The Concept of a Bacterium." *Archives of Microbiology* 42 (1962): 17–35.

van Niel, C. B. "The Classification and Natural Relationships of Bacteria." *Cold Spring Harbor Symposium on Quantitative Biology* 11 (1946): 285–301.

Whittaker, R. H. "New Concepts of Kingdoms of Organisms." *Science* 163 (1969): 150–160.

Woese, C. R., O. Kandler, and M. L. Wheelis. "Towards a Natural System of Organisms: Proposal for the Domains Archaea, Bacteria, and Eucarya." *Proceedings of the National Academy of Sciences USA* 87 (1990): 4576–4579.

Woese, Carl R. "Default Taxonomy: Ernst Mayr's View of the Microbial World." *Proceedings of the National Academy of Sciences USA* 95 (1998): 11043–11046.

Zimmer, Carl. "Scientists Unveil New 'Tree of Life.'" *New York Times*, April 12, 2016.

6. GENES FROM NEIGHBORS

Avery, O. T., C. M. Macleod, and M. McCarty. "Studies on the Chemical Nature of the Substance Inducing Transformation of Pneumococcal Types: Induction of Transformation by a Desoxyribonucleic Acid Fraction Isolated from Pneumococcus Type III." *Journal of Experimental Medicine* 79 (1944): 137–158.

Cohen, Stanley N., Annie C. Y. Chang, and Leslie Hsu. "Nonchromosomal Antibiotic Resistance in Bacteria: Genetic Transformation of *Escherichia coli* by R-Factor DNA." *Proceedings of the National Academy of Sciences USA* 69 (1972): 2110–2114.

Davis, Bernard. "Nonfilterability of the Agents of Genetic Recombination in *Escherichia coli*." *Journal of Bacteriology* 60 (1950): 507–508.

de Kruif, Paul. *The Microbe Hunters*. Orlando: Harcourt, 1926.

Griffith, F. "The Significance of Pneumococcal Types." *Journal of Hygiene* 27, no. 2 (1928): 113–159.

Hayes, W. "Recombination of *Bact. coli K* 12: Unidirectional Transfer of Genetic Material." *Nature* 169 (1952): 118–119.

Mandel, H. and H. Higa. "Calcium-Dependent Bacteriophage DNA Infection." *Journal of Molecular Biology* 53 (1970): 159–162.

Neidhardt, Frederick, John L. Ingraham, and Moselio Schaechter. *Physiology of the Bacterial Cell*. Sunderland, MA: Sinauer Associates, 1990.

Pollock, M. R. "The Discovery of DNA: An Ironic Tale of Chance, Prejudice and Insight. Third Griffith Memorial Lecture." *Journal of General Microbiology* 63 (1970): 1–20.

Tatum, E. L. and Joshua Lederberg. "Gene Recombination in the Bacterium *Escherichia coli.*" *Journal of Bacteriology* 53 (1947): 673–684.

Watanabe, Tsutomu. "Infective Heredity of Multiple Drug Resistance in Bacteria." *Bacteriological Reviews* 27 (1963): 87–115.

Watson, James D. *The Double Helix: A Personal Account of the Discovery of the Structure of DNA.* New York: Scribner, 1968.

Zinder, Norton D. and Joshua Lederberg. "Genetic Exchange in Salmonella." *Journal of Bacteriology* 64 (1952): 679–699.

7. CAN THE RECEIVING CELL SAY NO?

Baltimore, D., P. Berg, M. Botchan, D. Carroll, R. A. Charo, G. Church, J. E. Corn, G. Q. Daley, J. A. Doudna, M. Fenner, H. T. Greeley, M. Jinek, G. S. Martin, E. Penhoet, J. Puck, S. H. Sternberg, J. S. Weissman, and K. R. Yamamoto. "A Prudent Path Forward for Genomic Engineering and Germline Gene Modification." *Science* 348, no. 6230 (2015): 36–38.

Bertani, G. and J. J. Weigle. "Host Controlled Variation in Bacterial Viruses." *Journal of Bacteriology* 65 (1953): 113–121.

Bertani, Giuseppe. "Lysogeny at Mid-Twentieth Century: P1, P2, and Other Experimental Systems." *Journal of Bacteriology* 186 (2004): 595–600.

Lander, Eric S. "The Heroes of CRISPR." *Cell* 16 (2016): 18–28.

Lawrence, Jeffery G. and Howard Ochman. "Molecular Archeology of the *Escherichia coli* Genome." *Proceedings of the National Academy of Sciences USA* 95 (1998): 9413–9417.

Wade, Nicholas. "Scientists Seek Moratorium on Edits to Human Genome that Could Be Inherited." *New York Times,* December 3, 2015.

8. CAN THE TREE BE TRUSTED?

Cracraft, Joel and Michael J. Donoghue. *Assembling the Tree of Life.* New York: Oxford University Press, 2004.

Gest, H. "Samuel Ruben's Contributions to Research on Photosynthesis and Bacterial Metabolism with Radioactive Carbon." *Photosynthesis Research* 80 (2004): 77–83.

Gray, M. W. and W. F. Doolittle. "Has the Endosymbiont Hypothesis Been Proven?" *Microbiological Reviews* 46 (1982): 1–42.

Margulis, L. *Origin of Eukaryotic Cells.* New Haven, CT: Yale University Press, 1970.

Margulis, Lynn. *Once Upon a Time.* Barcelona: Editorial Septimus, 2013.

Margulis, Lynn and Dorion Sagan. *Microcosmos: Four Billion Years of Microbial Evolution.* Berkeley: University of California Press, 1986.

Sagan, Dorion, ed. *Lynn Margulis: The Life of a Scientific Rebel.* A Sciencewriters Book. White River Junction, VT: Chelsea Green Publishing, 2012.

Sagan, Lynn. "On the Origin of Mitosing Cells." *Journal of Theoretical Biology* 14 (1967): 225–274.

Sapp, Jan. "Too Fantastic for Polite Society: A Brief History of Symbiosis Theory." In *Lynn Margulis: The Life of a Scientific Rebel,* ed. Dorion Sagan. A Sciencewriters Book. White River Junction, VT: Chelsea Green Publishing, 2012, pp. 66–67.

Schopf, J. William. *Cradle of Life: The Discovery of Earth's Earliest Fossils: Impact of Horizontal Gene Transfer on the Tree of Life.* Princeton, NJ: Princeton University Press, 1999.

Schopf, J. William, ed. *Life's Origins: The Beginnings of Biological Evolution.* Berkeley: University of California Press, 2002.

Spang, Anja, Jimmy H. Saw, Steffen L. Jørgensen, Katarzyna Zaremba-Niedzwiedzka, Joran Martijn, Anders E. Lind, Roel van Eijk, Christa Schleper, Lionel Guy, and Thijs J. G. Ettema. "Complex Archaea that Bridge the Gap between Prokaryotes and Eukaryotes." *Nature* 521, no. 7551 (2015): 173–179.

Zablen, L. B., M. S. Kissil, C. R. Woese, and D. E. Buetow. "Phylogenetic Origin of the Chloroplast and Prokaryotic Nature of Its Ribosomal RNA." *Proceedings of the National Academy of Sciences USA* 72 (1975): 2418–2422.

Zimmer, Carl. "*Wolbachia:* A Tale of Sex and Survival." *Science* 292 (2001): 1093–1095.

9. THE TREE'S ECOLOGICAL FRUIT

DiGiulio, Daniel B., Benjamin J. Callahan, Paul J. McMurdie, Elizabeth K. Costello, Deirdre J. Lyell, Anna Rogaczewska, Christine L. Sun, Daniela S. A. Goltsman, Ronald J. Wong, Gary Shaw, David K. Stevenson, Susan P. Holmes, and David A. Relman. "Temporal and Spatial Variation of the

Human Microbiota during Pregnancy." *Proceedings of the National Academy of Sciences USA* 112 (2015): 11060–11065.

Eakin, Emily. "The Excrement Experiment: Treating Disease with Fecal Transplants." *New Yorker*, December 1, 2014.

Evans, Paul, Donovan H. Parks, Grayson L. Chadwick, Steven J. Robbins, Victoria J. Orphan, Suzanne D. Golding, and Gene W. Tyson. "Methane Metabolism in the Archaeal Phylum Bathyarchaeota Revealed by Genome-Centric Metagenomics. *Science* 350 (2015): 434–437.

Fierer, Noah, Christian L. Lauber, Nick Zhou, Daniel McDonald, Elizabeth K. Costello, and Rob Knight. "Forensic Identification Using Skin Bacterial Communities." *Proceedings of the National Academy of Sciences USA* 107 (2010): 6477–6481.

Vrieze, Anne, Els Van Nood, Frits Holleman, Jarkko Salojärvi, Ruud S. Kootte, Joep F. W. M. Bartelsman, Geesje M. Dallinga-Thie, Mariette T. Ackermans, Mireille J. Serlie, Raish Oozeer, Muriel Derrien, Anne Druesne, Johan E. T. Van Hylckama Vlieg, Vincent W. Bloks, Albert K. Groen, Hans G. H. J. Heilig, Erwin G. Zoetendal, Erik S. Stroes, Willem M. de Vos, Joost B. L. Hoekstra, and Max Nieuwdorp. "Transfer of Intestinal Microbiota from Lean Donors Increases Insulin Sensitivity in Individuals with Metabolic Syndrome." *Gastroenterology* 143, no. 4 (2012): 913–916.

10. THE TREE'S BEGINNINGS

Check, T. R., A. J. Zaug, and P. J. Grabowski. "In Vitro Splicing of the Ribosomal RNA Precursor of *Tetrahymena:* Involvement of a Guanosine Nucleotide in the Excision of the Intervening Sequence." *Cell* 27 (1981): 487–496.

Crick, F. H. C. and L. E. Orgel. "Directed Panspermia." *Icarus* 19 (1973): 341–346.

Deamer, David. *First Life: Discovering the Connection between Stars and How Life Begins.* Berkeley: University of California Press, 2011.

Dyson, Freeman. *Origins of Life.* Cambridge: Cambridge University Press, 1999.

Fry, Iris. *Emergence of Life on Earth: A Historical and Scientific Overview.* Piscataway, NJ: Rutgers University Press, 2000.

Gilbert, Walter. "Origin of Life: The RNA World." *Nature* 319 (1986): 618.

Lane, Nick, John F. Allen, and William Martin. "How Did LUCA Make a Living? Chemiosmosis in the Origin of Life." *BioEssays* 32, no. 4 (2010): 271–280.

Martin, William and Michael J. Russell. "On the Origins of Cells: A Hypothesis for the Evolutionary Transition from Abiotic Geochemistry to Chemoautotrophic Prokaryotes, and from Prokaryotes to Nucleated Cells." *Philosophical Transactions of the Royal Society of London B: Biological Sciences* 358 (2003): 59–85.

Miller, Stanley L. "Production of Amino Acids under Possible Primitive Earth Conditions." *Science* 117 (1953): 528–529.

Miller, Stanley L. and Harold C. Urey. "Organic Compound Synthesis on the Primitive Earth." *Science* 130 (1959): 245–251.

Mitchell, P. "Coupling of Phosphorylation to Electron and Hydrogen Transfer by Chemi-Osmotic Type of Mechanism." *Nature* 191 (1961): 144–148.

Patel, Bhavesh H., Claudia Percivalle, Dougal J. Ritson, Colm D. Duffy, and John D. Sutherland. "Common Origins of RNA, Protein and Lipid Precursors in a Cyanosulfidic Protometabolism." *Nature Chemistry* 7 (2015): 301–307.

Powner, Matthew W., Béatrice Gerland, and John D. Sutherland. "Synthesis of Activated Pyrimidine Ribonucleotides in Prebiotically Plausible Conditions." *Nature* 459 (2009): 239–242.

Wolman, Yecheskel, William J. Haverland, and Stanley L. Miller. "Nonprotein Amino Acids from Spark Discharges and Their Comparison with the Murchison Meteorite Amino Acids." *Proceedings of the National Academy of Sciences USA* 69 (1972): 809–811.

Acknowledgments

I am grateful to Catherine Vigran for standing by, ready to take over; to Anna Vigran and Michael Rieger for invaluable help obtaining illustrations; to Elena Ingraham for composing certain illustrations; and to Nancy O'Connor, Emilio Garcia, Roberto Kolter, Norman Pace, and Frank Harold for reading the manuscript and offering valuable advice.

Illustration Credits

Figure 1. Charles Darwin, *On the Origin of Species by Means of Natural Selection, or the Preservation of Favoured Races in the Struggle for Life.* London: John Murray, 1859. Wellcome Library, London. L0067068 (CC-BY-4.0).

Figure 2. Darwin's notebook. Archives and Manuscripts, Wellcome Library, London. L0003799 (CC-BY-4.0).

Figure 5. Swan-necked flask. Wellcome Library, London. M0012521 (CC-BY-4.0).

Figure 6. *Vigna unguiculata* nodules. Photo © Harry Rose / Flickr (CC-BY-4.0).

Figure 8. Redrawn by permission of the authors from George E. Fox, Linda J. Magrum, William E. Balch, Ralph S. Wolfe, and Carl E. Woese, "Classification of Methanogenic Bacteria by 16S Ribosomal RNA Characterization." *Proceedings of the National Academy of Sciences USA* 74, no. 10 (1977):4537–4541. Figure 1.

Figure 9. Ernst Heinrich Haeckel. *Generelle Morphologie der Organismen: Allgemeine Grundzüge der Organischen Formen-Wissenschaft, Mechanisch Begründet Durch die von Charles Darwin Reformirte Descendenztheorie* (Berlin: G. Reimer, 1866). Plate 1, Taf. 1. Wellcome Library, London. L0078506 (CC-BY-4.0).

Figure 10. Electron microscope facility, Dartmouth College. Prepared and imaged by Louisa Howard, Microscopist.

Figure 11. Electron micrograph image by Charles C. Brinton Jr., published in *TIME Magazine,* 109, no. 16 (April 18, 1977).

Figure 12. Bilayer scheme. Mariana Ruiz Villarreal *(LadyofHats),* Wikimedia Commons.

Index

Note: Page references in *italics* indicate figures.

Acetic acids, 251
Achman, Howard, 180
Acraea encendana, 211
Adamchak, Raoul, 43
Adelberg, Edward, 102
Adenine, 62, 72, 81
Adenine-thymine bonds, 83
Aerobic respiration, 54; energy (ATP) generation in, 198–199, 247–250; evolutionary opportunities in, 199; metabolic advantages of, 198–199; prokaryotic origins of, 189–190
African butterfly, *Wolbachia* in, 211
Agriculture: Green Revolution in, 42–44; Haber-Bosch process and, 41–44; organic, 43–44
Agrobacterium tumefaciens, 187–189
Aigarchaeota, 206
Alabaster Coast of Normandy, 25
Albatrosses, plastics and, 55
Albion (Great Britain), 25
Albumin. *See* Serum albumin
Alcoholic beverages, 22–23
Algae, 21, 24; bacterial origins of, 201, 203; blue-green (*see* Cyanobacteria); red, chloroplasts of, 204; red, digestion of, 227; similarity grouping of, 52; on Tree of Life, *7, 9.* *See also* Protists
Allen, John, 247, 250–251
Allomyces macrogynus, 155
Alpha-proteobacteria, 205
Altman, Sidney, 244

Amann, Rudolf, 230
Amino acid(s), 60–61; activated forms of, 90, 91–92; endosymbionts supplying, 208–210; formation, in origin of life, 239–240, 251; triplet codes for, 86–88, 89–90
Amino acid sequences, 16–17, 61–73; Crick on, 61; cytochrome c, 73; evolutionary history in, 61–73; hemoglobin, 63–67; Pauling's study of, 61–67, 72, 73; relatedness based on, 71–73; as Rosetta Stone of evolution, 64; Sanger's work on, 97–99
Aminoacyl-tRNA, 90
Aminoacyl-tRNA synthetases, 179
Amitochondrial eukaryotes, 205–206
Ammonia, 36, 39–44; anaerobic oxidation of, 40–41, 231; formation in Haber-Bosch process, 41–44; in origin of life, 237
Amoebic dysentery, 24
Amphipathic substances, 240–241, *241*
Amplification, DNA, 133–134, 225
Amylase, 23
Anaerobic respiration, 57
Anammox, 39, 40–41, 231
Anaxagoras, 233
Animal(s): bacterial gene transfer to, 187; kingdom, 8–9, 115, 120, 121, *122, 123;* nomenclature for, 119; origin of, 116; on Tree of Life, *7,* 21, 116

INDEX

Antarctica, subglacial lakes of, 48
Anthropomorphism, 50
Antibiotic(s): iChip production of, 45–46; microbiome impact of, 221–223; resistance of, 44–47, 161–162, 164–166, 222
Antibiotic-associated diarrhea, 221–223
Antibody cross-reactivity, 68–69
Anticodon, 92
Antifungal drugs, 24
Antiviral drugs, 34
Aphids, endosymbiosis in, 208–209
Archaea, 21, 27–30; archaebacteria name change to, 113–114; bacterial gene transfer to, 187; bacteria vs., 29–30; in *Bergey's Manual,* 119; cellular architecture of, 21, 29; classification problem with, 52; discovery of, 2–3, 27, 29, 49, 106–115, 130; eukaryotes as sister group of, 118; extremophilic, 27–29, 113–114, 134; flagella of, 29; kinship with, 27; Mayr's praise and doubt on, 112–113; methanogens, 57, 106–115; as "missing link" from prokaryotes to eukaryotes, 206–208; new discoveries and expansion, 114, 118; nitrogen-fixing, 36; in ocean, 27–29, 134; paralogous enzymes of, 117–118; as percentage of world's biomass, 114; phyla of, 135–136, 218; plasma membrane of, 29; ribosomes of, 77, 79; scientific reaction / skepticism to, 109–110; TACK superphylum, 206; taxonomic principle of balance vs., 111–114; taxonomy of, 118–119; on Tree of Life, 7, 9, 30, 49, 135–136; Woese's press release and publicity on, 109–110

Archaebacteria: name change to archaea, 113–114; Woese's discovery of, 110–115
Arginine, 66
Aristotle, 9–12, 50; belief in spontaneous generation, 12; hemoglobin study of, 65; ladder of nature, 9–11, *10;* taxonomy of, 51
Arora, David, 49–50
Aspergillus oryzae, 23
Atomic weight, of isotopes, 193
ATP (adenosine triphosphate), 117, 198–199, 247–250
ATPase: in chemiosmotic theory, 249–250; at root of Tree of Life, 117–118
Australia: ferric oxide in, 197; fossil record in, 191–193; Murchison meteorite in, 240
Autotrophy, 192
Avery, Oswald, 82, 146–148, 150, 154
Azotobacter, 37

Bacillary dysentery, 164–166
Bacillus: electron microscopy of, 124–125; Woese's investigation of, 103
Bacillus megaterium, 124–125
Bacteria, 21, 30–31; archaea vs., 29–30; benefits from, 30; in *Bergey's Manual,* 118–119; biochemically based tests of, 58–60; "bug sorting," 59, 103, 118; candidate phyla radiation, 136–137; cellular architecture of, 21, 29, 124–125, 126, 129; challenge of identifying, 59–60; classification in plant kingdom, 8–9; classification problem with, 52–60; colors of, 53;

INDEX

cultivation and isolation of, 31; dietary needs of, 58; diversity of, 30; DNA studies of, 31; domain, 3, 9, 114–115; electron microscopy of, 124–126, 129; evolutionary innovations of, 30–31; extract of, 76; flagella of, 29; gene transfer from (cross-kingdom), 187–214; gene transfer in, 96, 141–169; Gram-negative, 31, 190, 218; Gram-positive, 46, 190; Haeckel's Monera, 121–123; kinship with, 30; metabolism of, 53–54, 57–59; nitrogen-fixing, 36–39; nucleoid of, 126; paralogous enzymes of, 117–118; photosynthetic, 30–31, 57; phyla of, 218; protists *vs.*, 129; ribosomes of, 77, 79; shapes of, 60; size of, 53; Stanier and van Niel's concept of, 126–130; taxonomy of, 118–119, 124–132; transformation in, 96, 141–153; on Tree of Life, 7, 9, 30, 31, 49, 135–137; viruses *vs.*, 128–129; Woese's classification of, 110, 218

"Bacterial Evolution" (Woese), 114
Bacterial viruses. *See* Bacteriophages
Bacteriology, 9
Bacteriophages, 33, 85–87; defenses against, 170–177; electron microscopy of, 148, *149;* gene transfer in, 148–150, 167–168; Hershey-Chase experiment using, 148–150; lambda, 152, 171; virion of, 148, 168
Bacteroides: in gut microbiome / obesity, 227–229; Woese's classification of, 218
Bacteroides plebeius, 227
Bacteroids, 38
Baker, Richard, 124–125

Baker's yeast. *See Saccharomyces cerevisiae*
Balance, taxonomic principle of, 111–114
Balch, William E., 106–107
Banfield, Jillian F., 136–137
Barnacles, Darwin's study of, ix
Barnett, Leslie, 85–88
Basal radiation, 135
Bassham, James, 195–196
Baumann, Paul, 208–209
Baumannia cicadellinicola, 209–210
Beadle, George W., 154–155
Beermaking, 23
Beijerinck, Martinus, 33, 127
Benson, Andrew, 195–196
Bergey's Manual of Systematic Bacteriology, 118–119
Bertani, Giuseppe, 170–172
Bible: spontaneous generation in, 12; taxonomy in, 7; Tree of Life in, 4
Binomial taxonomy, 51, 128
Biochemical genetics, 155
Biochemical mutants, 156–157
Biochemistry, cracking genetic code in, 84–85, 88
Biodegradable plastics, 56
Bioinformatics, 173
Biological Species Concept, 52
Biotin, in gene transfer experiment, 157–158
Bishop, David, 99
Bitter Springs fossils (Australia), 191–192
Black smokers, 27–28, 246–247, 250–251
"Bloom," on grapes, 22
Blue-green algae. *See* Cyanobacteria
Blue-green sharpshooter, 210
Bonen, Linda, 204
Bonobos, relatedness of, 69

Borlaug, Norman, 42
Bosch, Carl, 41
Botany, 9
Brain, energy metabolism of, 199
Brenner, Sydney, 85–88
Broad-spectrum antibiotics, microbiome impact of, 221–223
Bryant, Marvin, 105
B strain of *E. coli*, 159–160, 162
Buchnera, 208–209
"Bug sorting," 59, 103, 118
Bundling-board experiment, 158–159, 167

Calcium carbonate exoskeletons, 25–26
California condor, 56
Calvin cycle, 195–196
Candida albicans, 222
Candidate phyla radiation (CPR), 136–137
Capsid, 32, 34
Carbohydrates, 62
Carbon-12 (^{12}C), 193, 196–197
Carbon-13 (^{13}C), 193, 196–197
Carbon-14 (^{14}C), 193–196
Carbon dating, 193–197
Carbon dioxide: bacteria metabolism of, 57; fixation by protists, 24–25; fixation during photosynthesis, 195–197; metabolism based on, 192; in origin of life, 237, 239, 251; release in fermentation, 22–23; repository in chalk, 25–26
Carbonic acid, 25–26
Carbon isotopes, 193–197
Carboxylic acid, 240–241
cas (CRISPR-associated sequences), 174
Cas9 nuclease, 174–177
Catalysts, 62, 65

Cattle, methane from, 104–105
Cavendish Laboratory (England), 85–88
Cech, Thomas, 244
Cell membrane: archaea, 29; chemiosmotic theory and, 247–251, 250; formation, in origin in life, 240–242; semipermeability of, 242
Cellulose, 55
Cenarchaeum symbiosum, 134
Central dogma, 82, 89, 95, 242–243
Chalk, 25–26
Chamberland, Charles, 33
Chamberland filter, 33
Chang, Annie, 152
Charcot, Jean-Baptiste, 27
Chargaff, Erwin, 82–83
Chargaff's rules, 82–83
Charpentier, Emmanuelle, 175–176
Chase, Martha, 148–150, 159
Chatton, Edouard, 129–130
Cheese making, *E. coli* in, 181–182
Chemiosmotic theory, 247–251, 250
Chemoautotrophy, 57, 192
Chert, fossil record in, 191–193
Chimpanzees, relatedness of, 69–71
Chinese rice wine, 23
Chlamydiae, 218
Chloramphenicol, 165, 221
Chloroplasts: bacterial origins of, 35–36, 189–190, 199, 200–208; circular DNA in, 190, 202; Lokiarchaeota as "missing link" to, 206–208; properties similar to bacteria, 190, 202; ribosomes of, 202; Woese's rRNA study of, 204
Chromium, 233
Citrus X disease, 209
Class, 51, 119

Classification, 7–9, 49–52; Aristotle's ladder of nature, 9–11, *10;* balance principle in, 111–114; *Bergey's Manual of Systematic Bacteriology,* 118–119; Copeland's four kingdoms, 123; dichotomous, 7, 8–9, 111–112, 119–120; domain, 3, 9, 114–115; electron microscopy and, 124–126; evolution and, 5, 7–8; fungi, 123; Haeckel's three kingdoms, 121–123, *122;* instinctive approach in, 49–50; kingdom, 8–9; Linnaeus system of, 50–52; *The Microbial World,* 101–102, 114; molecular, ix–x, 16–17, 119; morphological, ix, 49; path to Tree of Life, 9–16; phylogeny, 5; problem with bacteria in, 52–60; reproduction-based, 50–51; similarity groupings in, 50–52; tripartite division in, 52; virus, 34–35; Whittaker's five kingdoms, 123–124; Woese's analysis and discovery in, 99–115; Woese's focus on, 99, 118; Woese's primary kingdoms or urkingdoms in, 110–115

"Classification of Methanogenic Bacteria by 16S Ribosomal RNA Characterization" (Fox, Magrum, Balch, Wolfe, and Woese), 108

Clinton, Bill, 203

Cloning, 152–153

Clostridium, 37

Clostridium difficile, 222–223

Coacervates, 237

Coccidioidomycosis, 24

Coccolithophores, 25–26

Codon(s), 84–93; discovery of, 86–87; encoding capacity of, 87–88; Khorana's experiments with, 90–92; Nirenberg's experiments with, 88–90; redundancy of, 88, 178–179; triplet code of, 86–88

Codon preference, 178–179

Coenzyme(s), 103

Coenzyme M, 103–104

Cohen, Stanley, 152

Coming of the Golden Age, The (Stent), 92

Competence, for transformation, 151

Complementary base pairing, 80–81

Complement fixation, 68–69

"Concept of a Bacterium, The" (Stanier and van Niel), 126–130

"Concept of a Virus, The" (Lwoff), 128–129

Conjugation, 154–166, 177; antibiotic resistance in, 161–162, 164–166; bundling-board experiment in, 158–159; donors (F^+) in, 161–164; in double-mutant strains, 157–158; *E. coli,* 155–164, *163; fertile vs.* infertile strains in, 159–160; Hayes's experiment on, 160–162; Lederberg's studies of, 154–161; one-way gene transfer in, 160–164; plasmids in, 162–166; recipients (F^-) in, 161–162; sex pilus for, 162, *163; Shigella,* 164–166

Connick, Robert, 238

Copeland, Herbert F., 123

Cost, of DNA sequencing, 218–219, 231

CPR. *See* Candidate phyla radiation

Crenarchaeota, 136, 206, 218

Cretaceous Period, coccolithophores of, 25–26

Crick, Francis: on amino acid sequence, 61; central dogma of, 82, 89, 95, 242–243; discovery of DNA structure, 61–62, 80–82; elucidation

INDEX

Crick, Francis *(continued)*
  of genetic code, 85–88; on origin of life, 233, 234; on RNA's catalytic capacities, 244; on translation, 92
CRISPR, 172–177; Chinese experiments in embryos, 176; discovery of, controversy over, 173; ethical and practical concerns over use, 175–177; gene editing with, 173, 174–177; knockout mutations from, 176; meeting convened (2015) on, 177; palindromic sequences of, 173; simplification of system, 175–176; specific RNA guidance of, 174–175; variations in, 176–177
CRISPR-associated sequences *(cas)*, 174
CRISPR-Cpf1, 177
Cross-reactivity, and relatedness, 68–69
Crown gall disease, 187–189
Cultures / cultivation: bacteria, 31; methanogen, 104–106; as nomenclature requirement, 218; prokaryote, limits of, 217–218; yeasts, 21
Cyanobacteria: carbon dating and, 197; chloroplast origins in, 189–190, 201; dietary needs of, 58; eukaryotic capture of, 189–190; ferric oxide as marker for, 197–198; fossil record of, 190–193; metabolism of, 192; photosynthesis by, 30–31, 57; photosynthesis / oxygen origins in, 189–199, 203; prokaryotic nature of, 129; Woese's classification of, 218
Cysteine, 90
Cytochrome c, 73
Cytosine, 62, 72, 81, 83, 178–180

Darwin, Charles, 1, 3–7; evolutionary theory of, ix, 3–7; Haeckel and, 120, 121; infusoria of, 3, 7; on species definition, 51–52; Tree of Life concept of, 3, 3–7, 6; warm-little-pond hypothesis of, 235–240, 252–253
Davis, Bernard, 158–159, 167
Dayhoff, Margaret O., 116–117
DDT, 55, 56
Dead zone, 43
Deamer, David, 241–242
de Kruif, Paul, 154
Delbrück, Max, 150, 159
DeLong, Edward, 134
Denitrification, 39–41
Deoxyribonuclease (DNase), 147
Deoxyribonucleic acid. *See* DNA
Deoxyribose, 81
Descent with modification, 4, 186
Deuterium, 238
Diabetes mellitus, gut microbiome and, 228–229
Diarrhea: antibiotic-associated, 221–223; *E. coli* and, 183–185; *Giardia* and, 205–206
Diatomaceous earth (diatomite), 24–25
Diatoms, 24–25
Dichotomous classification, 8–9, 111–112, 119–120
Dideoxy method, 97–99
Digestion, microbiome's role in, 226–227
Diplomonads, 205–206
Directed panspermia, 234
Discovery: archaea, 2–3, 27, 29, 49, 106–115, 130; CRISPR, 173; DNA as tool for, 31, 48; DNA structure, 61–62, 80–84; macromolecules as tool of, 217–219; methanogen class, 219; rate of, 3; ribosome, 75–77; transformation, 141–146, 150; Tree of Life changes, 47–48; virus, 32–33

Disease: antibiotic-resistant, 44–47, 164–166; CRISPR and, 175–177; *E. coli* and, 182–185; fungal, 23–24; germ theory of, 33; pathogenicity islands and, 181–185; plant, endosymbiosis and, 209–210; plant, gene transfer and, 187–189; protist, 24; viral, 32–33

DNA, 16–17, 60; annealing of, 94–95; central dogma, 82, 242–243; Chargaff's rules on, 82–83; circular, in mitochondria and chloroplasts, 190, 202; complementary (specific) pairing in, 80–81; evolutionary history in, 62–63, 82; foreign, defenses against, 170–177; genetic information in, 62, 82–83; as genetic material, 142, 147–150, 154; Griffith's idioplasm as, 142, 147; Hershey-Chase experiment on, 148–150; hydrogen bonds of, 81, 94–95; modification of, 171–172; nucleobases of, 60–62, 72, 81–84; relatedness based on, 71–73; replication of, 81; separation / melting of, 94; structure of, 61–62, 80–84; as tool for discovery, 31, 48; transcription of, 91

DNA amplification, 133–134, 225

DNA hybridization, 94–96; DNA-DNA, 94–95; DNA-mRNA, 95–96

DNA sequencing, 133–137; advances in, 218–219; cost of, 218–219, 231; DeLong's archaea study, 134; evidence of gene transfer in, 177–180; online databases on, 135; PCR, 133–134, 225; Sanger's work on, 97–99; single-cell, 136; Venter's Sargasso Sea study, 134–135

DNA templates, 91

DNA viruses, 32, 34

Dogs, detection of human microbiome by, 226

Doi, Roy, 94–96, 99

Domains, 3, 9, 114–115, 119; controversy over, 119; gene transfer between, 187–214; kingdoms as subdivisions of, 114–115

Doolittle, W. Ford, 2, 204, 213

Doty, Paul, 94

Double Helix, The (Watson), 160–161

Double helix of DNA: annealing of, 94; complementary (specific) pairing in, 80–81; discovery of, 80–84; hybridization of, 94–96; separation of, 94

Double-mutant strains, 157–158

Doudna, Jennifer, 175–176

Doudoroff, Michael, 102

Dupetit, G., 39

Duplicated genes, at root of Tree of Life, 116–117

Dwarf wheat, 42

Dyson, Freeman, 251

E. coli. See *Escherichia coli*

Earth atmosphere: microbial impact on, 35–44; nitrogen fixation and resupply for, 35, 36–44; oxygen production for, 35–36; oxygen's appearance in, 190–199; primitive, and origin of life, 236–240

Ecology: coining of term, 120; microbial (*see* Microbial ecology)

Egyptians, fermentation by, 23

18S ribosomal RNA, 107

80S ribosomal RNA, 79

Einstein, Albert, 61

Eiseman, Ben, 222

Eisen, Michael, 173

Electron microscopy: bacteria, 124–126, 129; bacteriophage, 148, *149;* distortion during preparation, 125–126; ribosome, 74; virus, 33
Electrophoresis, 98–99
Electroporation, 152
Element abundance, as clue to origin of life, 233, 238
Elongation factor G, 118
Elongation factor Tu, 118
Endonucleases, restriction, 171–172
Endosymbionts, 208
Endosymbiosis: bacteria-insect, 208–212; chloroplast, 189–190, 200–206; Margulis on, 202–204; metabolic slaves in, 189, 199; mitochondria, 189–190, 200–206; nitrogen-fixing potential in, 212–213; obligate, 210
Energy transfer, chemiosmotic theory of, 247–251, *250*
Enteropathogenic *E. coli* (EPEC), 183
Enterotoxigenic *E. coli* (ETEC), 183
Enzymes: catalysts, 62, 65; cutting and analysis by, 65–67, 98–99, 100; genetic information in, 62; "one gene, one enzyme" principle for, 154–160; PaJaMo experiment, 75; paralogous, 117–118; RNA as (ribozymes), 243–244; viral, 32, 34
EPEC. *See* Enteropathogenic *E. coli*
Episome (plasmid), 162–164
Epulopiscium fishelsoni, 53
Escherichia coli: applications / uses of, 181–182; artificial transformation in, 152–153; cell membrane of, 242; conjugation in, 155–164, *163;* DNA of, 83; enzymes of, 65; fertile (K-12) vs. infertile (B) strains of, 159–160, 162; genetic code of, 89–90, 92–93; in gut and feces, 181; horizontal gene transfer in, 155–164, 180–185; as model organism, 22, 181; palindromic sequences in, 173; pathogenicity islands of, 181–185; pathogenic strains of, 182–185; percentage of transferred genes in, 180; phages infecting, 33, 85–87, 171; ribosomes of, 74; size of, 53; Tatum's collection of mutant strains of, 156
Escherichia coli O157:H7, 183–185
Essential amino acids, endosymbionts supplying, 208–210
ETEC. *See* Enterotoxigenic *E. coli*
Ettema, Thijs, 206
Eubacteria, 110–114
Euglena gracilis, 204
Eukaryotes: amitochondrial, 205–206; archaea as sister group of, 118; bacterial gene transfer to, 187–214; domain of, 3, 9, 114–115; evolution of, 189, 200–206; Lokiarchaeota as "missing link" to, 206–208; Margulis on, 202–204; nomenclature, 129–132; paralogous enzymes of, 117–118; phagocytosis by, 189–190; proto-, 204; ribosomes of, 77, 79; on Tree of Life, *7,* 9, 49; Woese's classification of, 110
Euryarchaeota, 136, 218, 219
Eutrophication, 43
Evans, Paul N., 219
Evolution: Aristotle and, 11; bacteria, 30–31; belief in, 4–5; Darwin's theory of, ix, 3–7; eukaryotic, 189, 200–206; gene transfer and, 177, 180, 185–186, 188–189, 213–214; history in DNA, 62–63, 82; history

276

INDEX

in macromolecules, 61–73; history in proteins, 61–73, 82; history in ribosomal RNA, 97; history in ribosomes, 73–79; history in RNA, 62–63, 74, 82; Huxley on coccolithophores, 26; intellectual leap on, 200; Lokiarchaeota as "missing link" in, 206–208; morphological studies of, ix; neo-Darwinian dogma on, 201–202; Rosetta Stone of, amino acid sequences as, 64; Rosetta Stone of, semantides as, 74; taxonomy and, 5, 7–8; virus, 35
Evolutionary clock, 69–71, 70
Exoskeletons, of coccolithophores, 25–26
Extremophiles, 27–29, 113–114, 134

F^+ (donor), in conjugation, 161–164
F^- (recipient), in conjugation, 161–162
Family, 51, 119
Famitsyn, Andrei, 201
Fatty acids: amphipathic nature, 240–241, *241;* membrane, and origin of life, 240–242
Fecal contamination, *E. coli* in, 181
Fecal microbiota transplantation (FMT), 223–224, 228–229
Feces, microbial population of, 220
Femaleness: bacterial, 161–162; insect, *Wolbachia* and, 211–212
Fermentation, 22–23, 57, 71
Ferric oxide (iron oxide), 57–58, 197–198
Filamentous fungi, 23
"Filterable viruses," 33
Fingerprint of microbes, human microbiome as, 225–226
Firmicutes, in gut microbiome / obesity, 227–229

FISH. *See* Fluorescence *in situ* hybridization
Fisher, R. A., 237
Fitch, Walter, 73
5S ribosomal RNA, 93, 98, 99
Fixation: ammonia, 36; carbon dioxide, 24–25, 195–197; nitrogen, 36–39, 41–44, 212–213, 231
Flagella, 29, 249
Flavobacteria, 218
Fluorescence *in situ* hybridization (FISH), 217, 229–231
FMT. *See* Fecal microbiota transplantation
Food and Drug Administration, on fecal transplantation, 224
Forensics, human microbiome in, 225–226
Fossil record, of cyanobacteria / oxygen production, 190–193
Fox, George E., 101–115; archaea discovery, 106–115; similarity index of, 101, 107–108, 111
F plasmid, 162–164
Franklin, Rosalind, 61–62
"Frozen accident theory," 93
Fructose, 22–23
Fungi, 21–24; bacterial gene transfer to, 187; classification in plant kingdom, 8–9, 123; classification problem with, 123; disease-causing, 23–24; filamentous, 23; kingdom, 115, 123–124; kinship with, 21; similarity grouping of, 52; on Tree of Life, 7, 9, 21; ubiquity of, 23

Gable, Max, 111
Gaia hypothesis, 203–204
Galactose, 59
Gayon, Ulysse, 39

277

Gene(s): duplicated, at root of Tree of Life, 116–117; exchange of (*see* Gene transfer); "genes on the loose," viruses as, 35; kinship-unique, 179–180; paralogous, 116–117
Gene editing, CRISPR for, 173, 174–177
Generalized transduction, 168
Genesis (biblical book), 4, 12
Genetically modified organisms (GMOs), 43–44
Genetic code, 72, 84–93; Cavendish quartet's experiments in, 85–88; climax of cracking, impact on science, 92; *E. coli*, 89–90, 92–93; Khorana's experiments in, 90–92; Nirenberg's experiments in, 88–90; redundancy in, 88, 178–179; syntax of, elucidation of, 85–88; triplet code in, 86–88; universality of, 92–93
Genetic Code: The Molecular Basis for Genetic Expression, The (Woese), 244–245
Genetic engineered plants, 188
Genetics: biochemical, 155; classical, 84; cracking genetic code in, 84–85; population, 237
Gene transfer, 141–169; antibiotic resistance in, 161–162, 164–166; Avery's study of, 146–148; bacteria to eukaryotes, 187–214; bacteria to insect, 208–212; in bacteriophages, 148–150, 167–168; bundling-board experiment in, 158–159, 167; CRISPR and immunity to, 172–177; damage from, 170; detection and defenses against, 169, 170–177; DNA as genetic material in, 142, 147–150, 154; in double-mutant strains, 157–158; *E. coli*, 155–166, 180–185; evidence of, 177–186; evolutionary significance of, 177, 180, 185–186, 188–189, 213–214; frequency of, studies of, 180–186; G plus C content in, 178–180; Griffith's studies of, 141–146; Hayes's experiment on, 160–162; Hershey-Chase experiment on, 148–150; intellectual leap on, 200; kinship-unique genes in, 179–180; Lederberg's studies of, 154–161, 167–168; in microbial sex (conjugation), 154–166, 177; mitochondria/chloroplast origins in, 189–190, 199, 200–208; pathogenicity islands in, 180–185; phagocytosis and, 189–190; plant disease from, 187–189; plasmid-mediated, 162–166; prokaryotic phylogeny *vs.*, 186; promiscuity in, 187; recombinant DNA technology, 152–153; restriction/modification system and, 170–172; selective advantages of, 151–152, 169; *Shigella*, 164–166; in transformation, 96, 141–153, 177; viral (transduction), 167–169, 177
Genome, 83; evidence of gene transfer in, 177–180; human, 97, 221
Genomes Online Database, 180
Genus, 51, 119
Gerland, Beatrice, 245
"Germ-free" mice, 227
Germ theory of disease, 33
Giardia lamblia, 205–206
Gibbons, relatedness of, 69
Gilbert, Walter, 245
Glassy-winged sharpshooter, 209–210
Globigerina, 26
Glucose, 59; fermentation, 22–23, 198; gut microbiome and levels of, 228–229; lactose metabolism to, 59; metabolism, energy from, 198–199

Glycine, 90
GMOs. *See* Genetically modified organisms
GOE. *See* Great Oxygen Event
Gogarten, Johan Peter, 117
Golden rice, 43–44
Gorillas, relatedness of, 69–71
Gosling, Raymond, 62
G plus C content, as evidence of gene transfer, 178–180
Gram-negative bacteria, 31, 190
Gram-positive bacteria, 46, 190, 218
Grapes: Pierce's disease in, 209–210; winemaking from, 22–23, 58, 71
Gray, Michael, 204–205
Great Oxygen Event (GOE), 198
Greenland, fossil record in, 191
Green nonsulfur bacteria, 218
Green Revolution, 42–44
Green sulfur bacteria, 218
Griffith, W. Frederick, 141–146, 150
Growth / no growth experiments, 156
Guanine, 62, 72, 81, 83, 178–180
Gulf of Mexico, dead zone in, 43
Gut microbiome, 226–229; antibiotics and, 221–223; Crohn's disease and, 223–224; diabetes mellitus and, 228–229; fecal transplant and, 223–224, 228–229; obesity and, 227–229
Gyres, 54–55

Haber, Fritz, 41
Haber-Bosch process, 41–44
Hadean eon, 190–191, 232
Haeckel, Ernst, 120–123, 201
Haemophilus influenzae, transformation in, 151
Haldane, J. B. S., 80–81, 236–237
Haloferax mediterranei, 173
Halophiles, 28, 113–114, 173
Haloquadratum walsbyi, 28
Harold, Frank, 186
Hayes, William, 160–162
Hay infusions, 13–16
Helmholtz, Hermann von, 234
Hemoglobin: composition of, 63, 65; enzyme cutting and analysis of, 65–67; information content of, 63; limitations of studies, 72; molecular weight of, 75; Pauling's amino acid study of, 63–67, 71, 72
Hershey, Alfred, 148–150
Hershey-Chase experiment, 148–150
Heterotrophs, first organisms as, 236
Higa, Akiko, 152
High frequency of recombination (Hfr), 164
Histidine, in gene transfer experiment, 156–157
HIV. *See* Human immunodeficiency virus
Holley, Robert W., 91–92
Homalodisca vitripennis, 209–210
Horizontal gene transfer, 141–169; antibiotic resistance in, 161–162, 164–166; Avery's study of, 146–148; bacteria to eukaryotes, 187–214; bacteria to insect, 208–212; in bacteriophages, 148–150, 167–168; bundling-board experiment in, 158–159, 167; CRISPR and immunity to, 172–177; damage from, 170; detection and defenses against, 169, 170–177; DNA as genetic material in, 142, 147–150, 154; in double-mutant strains, 157–158; *E. coli*, 155–166, 180–185; evidence of, 177–186; evolutionary significance of, 177, 180, 185–186, 188–189,

Horizontal gene transfer *(continued)* 213–214; frequency of, studies of, 180–186; G plus C content in, 178–180; Griffith's studies of, 141–146; Hayes's experiment on, 160–162; Hershey-Chase experiment on, 148–150; intellectual leap on, 200; kinship-unique genes in, 179–180; Lederberg's studies of, 154–161, 167–168; in microbial sex (conjugation), 154–166, 177; mitochondria/chloroplast origins in, 189–190, 199, 200–208; pathogenicity islands in, 180–185; phagocytosis and, 189–190; plant disease from, 187–189; plasmid-mediated, 162–166; prokaryotic phylogeny *vs.*, 186; recombinant DNA technology, 152–153; restriction/modification system and, 170–172; selective advantages of, 151–152, 169; *Shigella,* 164–166; in transformation, 96, 141–153, 177; viral (transduction), 167–169, 177
Host-controlled variation, 171
Hsu, Leslie, 152
Huang, Junjiu, 176
Human genome, sequencing of, 97
Human Genome Project, 221
Human growth hormone, production of, 152–153
Human immunodeficiency virus (HIV), 33, 34, 243
Human microbiome, 214, 220–231; acquisition and development of, 224–225; antibiotics and, 221–223; *Candida albicans* in, 222; care and feeding of, 221–226; Crohn's disease and, 223–224; diabetes mellitus and, 228–229; digestive role of, 226–227; fecal, 220; fecal transplantation and, 223–224, 228–229; FISH analysis of, 229–231; forensic value of, 225–226; gut, 226–229; individual variation and distinctiveness of, 221, 225–226; obesity and, 227–229; odor from, 226; population of, 220–221; premature birth and, 229
Human Microbiome Project, 220–221
Hungate, Robert E., 102, 104–106
Hungate Technique, 104–106
Huxley, Thomas Henry, 26
Hybridization, 94–96; DNA-DNA, 94–95; DNA-mRNA, 95–96; FISH, 217, 229–231; relatedness revealed in, 95
Hydrocarbons, 241
Hydrogen: isotopes of, 238; in origin of life, 237, 238, 251
Hydrogen bonds, DNA, 81; annealing of, 94–95; separation of, 94; similarities *vs.* differences in, 83
Hydrogen cyanide, in origin of life, 252
Hydrogen sulfide, 57, 192–193
Hydrophilic heads, 240–241, *241*
Hydrophobic tails, 240–241, *241*
Hydrothermal vents: black smokers, 27–28, 246–247, 250–251; origin of life at, 246–251; white smokers, 246–247
Hyperthermophilic protists, 28

iChip, 45–46
Idioplasm, 142, 147
Imperial Chemical Industries, 56
Index of dissimilarity (ID), 69–71
India, Green Revolution in, 42, 43–44
Infection thread, in legumes, 38

"Infective heredity," 166
Influenza virus, 243
Infusoria, 3, 7
Insects: endosymbiosis in, 208–212; gender, *Wolbachia* and, 211–212
Insulin: production of, 152–153; structure of, 97–98
Intron, 244
Iron oxide, 57–58, 197–198
Iron-reducing bacteria, 57
Isotopes: carbon, 193–197; definition of, 193
Isua sequence, 191
Ivanovski, Dimitri, 33
Iwabe, Naoyuki, 118

Jacob, François, 75
Jenner, Edward, 32
Jetten, Mike S. M., 40

Kamen, Martin, 193, 196
Kandler, Otto, 114–115, 120
Kellenberger, Eduard, 125–126
Khorana, Har Gobind, 90–92
Kingdoms, 8–9, 51, 120; Copeland's four, 123; gene transfer between, 187–214; Haeckel's three, 121–123, *122*; primary, Woese's, 110–115; prokaryote-eukaryote distinction, 129–132; as subdivisions of domains, 114–115; Whittaker's five, 123–124
Kinship, 1–3, 35, 47; with archaea, 27; with bacteria, 30; with fungi, 21; with primates, 69–71
Kinship-unique genes, 179–180
Kluyver, Albert J., 127
Knight, Rob, 225–226
Knockout mutations, CRISPR and, 176

Korarchaeota, 206
K-12 strains of *E. coli*, 159–160, 162

D-lactate hydrogenase, 65–66
D-lactic acid, 65
Lactic acid bacteria, 58
Lactose, ability to utilize, 58–59
Lactose intolerant, 59
Ladder of nature, Aristotle's, 9–11, *10*
Lake Vostoc, Antarctica, 48
Lambda phage, 152, 171
Lander, Eric, 173
Lane, Nick, 247, 250–251
Last universal common ancestor (LUCA), 245
Lawrence, Jeffrey, 180
Lederberg, Esther, 167–168
Lederberg, Joshua, 154–161, 167–168, 220, 234
Leeuwenhoek, Antonie van, 3, 30, 127, 205, 219
Leghemoglobin, 38–39
Legumes, nitrogen fixation in, 37–39, *38*, 213
Lewis, Kim, 45
Life. *See* Origin of life; Tree of Life
Lignin, 62
Linnaeus, Carl, 50–52, 120, 128
Lipases, 65
Lipids, 62
Liquor, distillation of, 23
Lobar pneumonia, 142–143
Loeb, Jacques, 253
Lokiarchaeota, 118, 206–208
Lord Kelvin (William Thomson), 234
Lovelock, James, 203
LUCA (last universal common ancestor), 245
Luria, Salvador, 110, 150

Lwoff, André, 128–129
Lysenko, Trofim, 236–237
Lysine, 66, 89

MacLeod, Colin, 82, 147
Macrobes, 9
Macromolecules, 16–17, 49, 60–73; discovery and identification via, 217–219; evolutionary clock in, 69–71, *70;* evolutionary history in, 61–73; genetic information in, 62; Pauling's study of, 61–67, 71, 72; properties of, 60–61; relatedness based on, 71–73. *See also* DNA; Proteins; RNA
Maggots, spontaneous generation and, 12
Magrum, Linda J., 106–107
Malaria, 24, 83, 237
Maleness, bacterial, 161–164
Malic acid, 58, 71
Malt, 23
Maltose, 23
Mamavirus, 34
Mandel, Manley, 152
Margoliash, Emmanuel, 73
Margulis, Lynn, 123–124, 202–204
Marmur, Julius, 94, 96
Martin, William, 247, 250–251
Mayr, Ernst, 52, 112–113, 132
McCarty, Maclyn, 82, 147
Medawar, Peter, 32, 237
Meister, Joseph, 32–33
Membranes, cellular: archaea, 29; chemiosmotic theory and, 247–251, *250;* formation, in origin of life, 240–242; semipermeability of, 242
Mendel, Gregor, 84
Merozygote, 160–161
Mesosome, 125

Messenger RNA (mRNA), 82, 89–92; catalytic / enzyme capacities of, 243–244; hybridization with DNA, 95–96
Metabolic slaves, in endosymbiosis, 189, 199
Metabolix, 56
Metabolome, 220
Metagenomics, 229–231
Metascience, phylogeny as, 127–128
Meteorites: amino acids in, 240; building blocks of life from, 234–235, 240; lipid-like compounds in, 241–242; origin of life from, 233–235
Methane, 57, 102, 103; cattle as source of, 104–105; metabolism based on, 27, 57, 102–115 (*see also* Methanogens); in origin of life, 236, 238, 239, 251
Methanobacterium thermoautothrophicum, 106, 109
Methanogens, 57, 102–115; archaea discovery, 57, 106–115; cell walls of, 104; coenzyme M of, 103–104; culturing of, 104–106; discovery of new class, 219; phylogenetic tree of, *107,* 107–108; unique biology of, 103–104
Methicillin-resistant *Staphylococcus aureus* (MRSA), 45–47
Methionine, 88, 156–158
Mexico, Green Revolution in, 42
Mice, microbiome studies in, 226–228
Micheli, Pier Antonio, 12–13
Michigan State University, online database of, 135
Microbe(s): atmospheric effects of, 35–44; beginning of Tree of Life, 6–7; Darwin's infusoria, 3, 7;

INDEX

dominance on Tree of Life, 9, 21; impact of, 35–47. *See also specific microbes and types of microbes*
Microbe Hunters, The (de Kruif), 154
Microbial ecology: definition of, 217; DeLong's archaea study, 134; FISH analysis of, 217, 229–231; increased interest in, 44; Tree of Life and, 217–219; Venter's Sargasso study, 134–135
"Microbial infallibility," 54
Microbial World, The (Stanier, Ingraham, and Wheelis), 101–102, 114
Microbiological Reviews, 203
Microbiome, 9, 214; definition of, 220; human, 214, 220–231; mice studies of, 226–228. *See also* Human microbiome
Microbiota, 9, 220. *See also* Human microbiome; Microbiome
Micrococcus, 83
Microflora, 9
"Microsomes," 77. *See also* Ribosomes
Microsporidia, 205
Midway Atoll, plastics and albatrosses on, 55
Miller, Stanley L., 237–240
Miller-Urey experiment, 237–240, 246
Mimivirus, 34
Minimal medium, 157
Minnesota, oxidized iron deposits in, 197
Minus mutations, 86–87
"Missing link," Lokiarchaeota as, 206–208
Mitchell, Peter, 247–251
Mitochondria: bacterial origins of, 123–124, 189–190, 199, 200–208; circular DNA in, 190, 202;

Lokiarchaeota as "missing link" to, 206–208; properties similar to bacteria, 190, 202; ribosomes of, 202; wheat, study of, 204–205
Model organisms, 22
Modification, DNA, 171–172
Mojica, Francisco, 173
Molds, 21
Molecular classification, ix–x, 16–17, 119
"Molecular Structure of Nucleic Acids: A Structure for Deoxyribose Nucleic Acid" (Watson and Crick), 80
Molybdenum, 233
Monera, 121–123
Monod, Jacques, 75
Møns Klint, Denmark, 25
Moore's Law, 218
Morphology: classification based on, ix, 49; limitations of study, ix; molecular, ix–x
Moyle, Jennifer, 247
MRSA (methicillin-resistant *Staphylococcus aureus*), 45–47
Mullis, Kary, 133
Murchison meteorite, 240
Mushrooms, 21, 24
Mutations: evolutionary role of, 63–64; knockout, CRISPR and, 176; rates of, 71
Mycoplasma, 83
Mycology, 9

Nanoarchaeota, 218
National Human Genome Research Institute, 218
National Institutes of Health, 220–221
Natural gas (methane), 57, 102, 103
Needham, John, 12–13

Nematodes, *Wolbachia* in, 212
Neo-Darwinian dogma, 201–202
Neufeld, Friedrich, 146
Neurospora crassa, 154–155
Nidus, 145
Niftrik, Laura van, 40
Nirenberg, Marshall, 88–90, 91
Nitrates, 36, 39–44
Nitrite, 36, 39–40
Nitrogenase, 36–39
Nitrogen cycle, 35–44
Nitrogen fixation, 36–39; bacteria-plant symbiosis in, 37–39, 38, 213; endosymbiotic potential for, 212–213; energy required for, 36–37; FISH analysis for, 231; human (Haber-Bosch process), 41–44; oxygen sensitivity of, 37
Nitrogen resupply, 36, 39–41
Nitrous oxide, 39–40
Nomenclature (naming): archaea, 113–114; cultivation as requirement for, 218; Haeckel's, 120–123; as intellectual contact sport, 120; Linnaeus, 51, 120, 128; major divisions of life, 119–124; prokaryote-eukaryote distinction, 129–132; ribosome, 78; transformation, 150
Noneukaryote, 131
North Pacific Gyre or "Garbage Patch," 54–55
Nucleases, 172, 174–177
Nucleic acids, structure of, 60–61. *See also* DNA; RNA
Nucleobases, 60–62, 72, 81–84; annealing of, 94–95; Chargaff's rules on, 82–83; complementary (specific) pairing of, 80–81; DNA, 60–62, 72, 81–84; hydrogen bonds between, 81, 83; RNA, 60–62, 72,

243, 245; separation of, 94; sequencing of, 97–100, 118, 132–137; triplet codes of, 86–88
Nucleoid, 126
Nucleus, protist, 24

Obesity, gut microbiome and, 227–229
Obligate endosymbiosis, 210
Oceans: archaea of, 27–29, 134; extremophiles of, 27–28; origin of life in, 246–251; photosynthetic protists of, 24–25; plastics/gyres in, 54–55; viruses in, 32
Oenococcus oeni, 71
Oleander leaf scorch, 209
Olives, *Xylella* infection in, 210
"On a Piece of Chalk" (Huxley), 26
Onchocerca ochengi, 212
"One gene, one enzyme" principle, 154–160
Online databases, 135
On the Generation of Animals (Aristotle), 12
"On the Origin of Mitosing Cells" (Margulis), 202–203
On the Origin of Species (Darwin), ix, 4–7, 51–52
Oparin, Alexander, 236–237, 238
Oparin-Haldane hypothesis, 236–237
"Optically pure" air, 15–16
Orangutans, relatedness of, 69–71
Order, 51, 119
Organic agriculture, 43–44
Orgel, Leslie, 234, 244
Origin of life, 232–253; abundance of elements as clue to, 233, 238; amino acid formation in, 239–240, 251; chemiosmotic theory and, 247–251, 250; Crick on, 233, 234; Dyson's "origins" proposal on, 251; Earth's

atmosphere and, 236–240; hydrogen cyanide and, 252; membrane formation in, 240–242; Miller-Urey experiment on, 237–240, 246; in ocean (hydrothermal vents), 246–251; Oparin-Haldane hypothesis of, 236–237; panspermia theory of, 232–235, 240; reducing environments for, 236–240, 246–251; RNA / RNA World in, 242–246, 250–251; warm-little-pond hypothesis of, 235–240, 251, 252–253

Origin of Life, The (Oparin), 236

Origins, fascination with, 1–3. *See also* Evolution; Origin of life

Origins of Life (Dyson), 251

Oxphos (oxidative phosphorylation), 248

Oxygen: appearance in atmosphere, 190–199; ferric oxide as marker for, 197–198; in fossil record, 190–193; metabolic advantages of, 198–199; methanogen intolerance of, 104–105; microbial production of, 35–36; nitrogenase sensitivity to, 37; production, prokaryotic origins of, 189–199, 203; toxicity in Great Oxygen Event, 198; utilization, prokaryotic origins of, 189–190, 203

Oxygen cycle, 190

"Oxygen holocaust," 198

Oxygen sumps, 197–198

Ozone, origins of, 198

Pace, Norman, 130–132, 135–136, 244

Painter, Page, 102

PaJaMo experiment, 75

Palindromic sequences, of CRISPR, 173

Panspermia, 232–235, 240

Paralogous enzymes, 117–118

Paralogous genes, 116–117

Pardee, Arthur, 75–76

Pasteur, Louis, 14–16, 22, 32–33

"Patchy conservation," of rRNA, 230

Pathogenicity islands, 180–185

Pauling, Linus, 61–67, 71–73, 82, 84, 89

Payne, William Jackson, 56

PCR. *See* Polymerase chain reaction

Pease, Daniel, 124–125

Penicillin, 221

Peptidoglycan, 62, 104

Permeases, 242

PHA-based plastics, 56

Phage Group, 159

Phages (bacteriophages), 33, 85–87; defenses against, 170–177; electron microscopy of, 148, *149;* gene transfer in, 148–150, 167–168; Hershey-Chase experiment using, 148–150; lambda, 152, 171; virion of, 148, 168

Phagocytosis, 189–190, 207

Phenylalanine, 89

Phony peach disease, 209

Phosgene, 194–195

Phosphate, radioactive, 98, 132

Phospholipids, in membranes, 241

Photosynthesis: by bacteria, 30–31, 57; carbon dating of, 196–197; carbon dioxide fixation during, 195–197; definition of, 192; forms other than oxygen-releasing, 192–193; by plants, 35–36; prokaryotic origins of, 189–199, 203; by protists, 24–26

Phyla (sing., phylum), 51, 119, 120; archaea, 135–136, 218; bacteria, 218

Phylogeny, 5, 120–121; gene transfer vs., 186; as metascience, 127–128; methanogen, *107,* 107–108; *The*

Phylogeny *(continued)*
 Microbial World, 101–102; Woese's analysis and discovery in, 99–115, 118; Woese's focus on, 99, 118; Woese's primary kingdoms or urkingdoms in, 110–115
Phylotype, 52
Pierce's disease, 209–210
Piezophilic archaea, 28
Pilbara Supergroup, 191
Planctomyces, 218
Plant(s): bacterial gene transfer to, 187–214; kingdom, 8–9, 115, 120, 121, 122, 123; Linnaeus classification of, 50–52; nomenclature for, 119; PHA-based plastics from, 56; photosynthesis by, 35–36; photosynthesis by, prokaryotic origins of, 189–190; on Tree of Life, 7, 9, 21
Plant disease: endosymbiosis and, 209–210; gene transfer and, 187–189
Plasma membrane. *See* Cell membrane
Plasmids, 152, 162–166; defenses against, 170–177; F plasmid, 162–164; Ti, 188
Plasmodium falciparum, 83
Plastics: biodegradable, 56; insolubility of, 55; invulnerability to microbes, 54–56; ocean-borne (gyres), 54–55
Plastids, 190. *See also* Chloroplasts
Plate tectonics, 26, 200
Plato, 8–9, 9
Pneumococcus. See *Streptococcus pneumoniae*
Polymerase chain reaction (PCR), 133–134, 225
Polypeptides, 89
Population genetics, 237

Porphyran, digestion of, 227
Porphyridium, 204
Pouchet, Félix, 13–15
Pourquoi Pas? (research vessel), 27
Powner, Matthew, 245
Premature birth, vaginal microbiota and, 229
"Primae specie," 51
Primary kingdoms, 110–115
Primates: evolutionary clock of, 69–71, *70;* index of dissimilarity, 69–71; semantides and relatedness of, 68–71
Probe, in FISH, 229
Proceedings of the National Academy of Sciences, 110, 112, 117
Proflavin, mutations from, 86–87
Prokaryote(s), 21; archaea *vs.*, 29–30; in *Bergey's Manual,* 118–119; cellular architecture of, 21, 29; cellular identity of, 213–214; CRISPR and immunity in, 172–177; culturing of, limits of, 217–218; gene transfer in, 141–169; gene transfer *vs.* phylogeny / taxonomy, 186, 213–214; nitrogen-fixing, 36–39; nomenclature, 129–132; transformation in, 96, 141–153
Prokaryote-eukaryote distinction, 129–132; advocates for expunging prokaryote as term, 130–131; dogma of, paralysis from, 131; Mayr's defense of, 132; Woese's objection to, 130, 131
Proline, 89–90, 157–158
Promiscuity, in gene transfer, 166, 187
Proteases, 34, 66–67
Proteins, 16–17, 60–73; central dogma, 82, 242–243; enzyme cutting and analysis of, 65–67; evolutionary

clock in, 69–71, *70;* evolutionary history in, 61–73, 82; genetic code for, 72; genetic information in, 62; molecular weight of, 75; Pauling's study of, 61–67, 71, 72; relatedness based on, 71–73; in ribosomes, 74, 93; RNA translation to, 82, 89, 91–92; structure of, 60–61; synthesis in ribosomes, 74, 78, 88; ultracentrifuge of, 75; viral, 32
Proteobacteria, 218
Proteome, 220
Protists, 21, 24–26, 115; bacteria *vs.*, 129; cellular architecture of, 24; diatoms, 24–25; diversity of, 24; Haeckel's classification of, 121–123; kingdom, 121, *122,* 123; nomenclature, 121; pathogenic, 24; photosynthetic, 24–26; ubiquity of, 24
Protocells, 242
Proto-eukaryotes, 205
Proton gradients, and energy transfer, 247–251, *250*
Protoplasm, 12
Protozoa, 21, 24; classification in animal kingdom, 9; similarity grouping of, 52; on Tree of Life, *7, 9*. *See also* Protists
Protozoology, 9
Pseudomonas stutzeri, transformation in, 151
Pseudo wild-type gene, 86–87
Psychrophilic archaea, 134
"Purifying selection," 64
Pyrococcus yayanosii, 27–28

Rabbit pathogenic *E. coli,* 183
Rabies vaccine, 32–33
Radioactive phosphate, 98, 132
Radio-resistant micrococci, 218

Rapid lysing, 85
Recombinant DNA technology, 152–153, 172
"Recombination in *Bact. coli* K12: Unidirectional Transfer of Genetic Material" (Hayes), 160
Red algae: chloroplasts of, 204; digestion of, 227
Redi, Francesco, 12
Red rust (ferric oxide), 197–198
Reducing environments, origin of life in: atmospheric, 236–240; micro-environments, 239–240; oceanic, 246–251
Religion: evolution and, 4–5; spontaneous generation and, 12, 13
Rennin, from *E. coli,* 181–182
Replication, DNA, 81
Replication, in origin of life: Dyson's dual proposal on, 251; RNA's capacity for, 244–246; warm-little-pond hypothesis of, 235–240
Reproduction-based classification, 50–51
Resistance, antibiotic, 44–47, 161–162, 164–166
Resistance transfer factor (RTF), 166
Respiration, 57–58; aerobic, 54, 189–190, 198–199, 247–250; anaerobic, 57; "microbial infallibility," 54
Restriction endonucleases, 171–172
Restriction/modification system, 170–172
Reverse transcriptase, 34
Rhizobium, 37–39, 213
Rhizobium radiobacter, 187–189
Ribonuclease P (RNase P), 244
Ribonuclease T_1, 98–99, 100, 132–133
Ribonucleic acid. *See* RNA

Ribonucleotides, 245–246
Ribose, 243, 245
Ribosomal Database Project, 135
Ribosomal RNA (rRNA): *Bacillus,* 103; Banfield's study of, 136; chloroplast, 204; evolutionary inertia and reach of, 93–96; FISH analysis of, 230; methanogen, 106–115; "patchy conservation" of, 230; as Rosetta Stone of evolution, 97; signatures of, 100, 106; Woese's analysis and discovery in, 99–115, 118
Ribosomes, 34–35, 73–79, 93; composition of, 74, 79, 93; discovery of, 75–77; electron microscopy of, 74; in mitochondria and chloroplasts, 202; naming of, 78; protein synthesis in, 74, 78, 88; scientific interest in, 77–79; subunits of, 77, 93; Tree of Life and, 73–74, 78–79; Woese's proposed mechanism of movement, 101, 108
Ribozymes, 244
Ribulose bisphosphate carboxylase (RuBisCO), 195
Richards, Robert J., 121
rII bacteriophage, 85–87
River blindness, 212
RNA, 16–17, 60; catalytic/enzyme capacities of, 243–244; central dogma, 82, 89, 242–243; CRISPR guidance by, 174–175; DNA transcription to, 91; evolutionary history in, 62–63, 74, 82; FISH analysis of, 217, 229–231; genetic information in, 62, 243; Khorana's synthesis of, 90–92; Nirenberg's synthesis of, 89–90; nucleobases of, 60–62, 72, 243, 245; origin of life via, 242–246, 250–251; relatedness based on, 71–73; in ribosomes, 74, 79, 93; sequencing of, Woese's method of, 99–100, 118, 132–133, 219; translation of, 82, 89, 91–92
RNA, messenger, 82, 89–92; catalytic/enzyme capacities of, 243–244; hybridization with DNA, 95–96
RNA, ribosomal. *See* Ribosomal RNA
RNA, transfer, 90, 91–92
RNA viruses, 32, 34, 243–244
"RNA World, The" (Gilbert), 245
Roberts, Richard B., 78
Roberts, Richard J., 43–44
Robinow, Carl, 125–126
Root hairs, of legumes, 37
Root nodule, of legumes, 38–39
Rosetta Stone of evolution: amino acid sequences as, 64; ribosomal RNA as, 97; semantides as, 74
Roth, John, 64
Rough (R) strains, of pneumococcus, 144–148
RTF. *See* Resistance transfer factor
Ruben, Samuel, 193–196
RuBisCo, 195
Rumen, culturing methanogens in, 104–105
Russell, Michael, 251
Ryan, Francis J., 154–155, 157, 159

Saccharification, 23
Saccharomyces cerevisiae, 22, 77, 198–199
Sagan, Carl, 111, 234
Sagan, Lynn. *See* Margulis, Lynn
Sake, 23
Salivary amylase, 23
Salk Institute, 91

INDEX

Salmonella typhimurium (enteritidis), 167
Salt-tolerant microbes (halophiles), 28, 113–114, 173
Sanger, Frederick, 97–99
San Joaquin Valley fever, 24
Sapp, Jan, 200
Sargasso Sea, microbial ecology of, 134–135
Sarich, Vincent, 68
Schachman, Howard, 75–76
Schimper, Andreas, 201
Schopf, J. William, 191–192
Schwartz, Robert M., 116–117
Scott, William, 141
Semantides, 61–79, 82, 84, 89, 97; cytochrome c as, 73; evolutionary clock of, 69–71, *70;* hemoglobin as, 63–67; proteins (amino acids) as, 61–73; ribosomes as, 73–79; RNA as, 74; Rosetta Stone of evolution, 74; serum albumin as, 68–71
Semipermeable membranes, 242
Serology, 143
Serum albumin: evolutionary history in, 68–71; index of dissimilarity, 69–71; relatedness based on, 71–73
70S ribosomal subunit, 79
Sex, microbial (conjugation), 154–166, 177; antibiotic resistance in, 161–162, 164–166; bundling-board experiment in, 158–159; donors (F⁺) in, 161–164; in double-mutant strains, 157–158; *E. coli,* 155–164, *163;* fertile *vs.* infertile strains in, 159–160; Hayes's experiment on, 160–162; Lederberg's studies of, 154–161; one-way gene transfer in, 160–164; plasmids in, 162–166, 166; recipients (F⁻) in, 161–162; *Shigella,* 164–166

Sex pilus, 162–163, *163*
Shiga-like toxin, 185
Shigella, 164–166, 185
Shiva, Vandana, 43
Siamangs, relatedness of, 69
Sickle-cell anemia, 237
Signatures, of rRNA, 100, 106
Silica, 24–25
Similarity groupings, 50–52
Similarity index (S_{AB}), 101, 107–108
Singer, Charles, 11
Sisquoc Formation, 25
16S ribosomal RNA: *Bacillus,* 103; methanogen, 106–108; similarity index of, 101; Woese's analysis of, 99–108, 132–133
Skin microbiome: Cesarean section and, 224; as fingerprint of microbes, 225–226; odor from, 226
Sleeping sickness, 24
Smallpox vaccine, 32
Small subunit RNA, 107, 111
Smith, John Maynard, 204
Smooth (S) strains, of pneumococcus, 144–148
Socrates, 8
Space: building blocks of life from, 234–235, 240, 252; hydrogen cyanide from, 252; lipid-like compounds from, 241–242; origin of life in (panspermia), 232–235, 240
Spallanzani, Lazzaro, 12–13
Species: in *Bergey's Manual,* 119; biological concept / definition of, 52, 112; dilemma in defining, 51–52; "primae specie," 51; prokaryotic, gene transfer *vs.,* 186
Specific (complementary) pairing, 80–81

289

Spiegelman, Sol, 95, 99
Spirochaetes/spirochetes, 203–204, 218
Spontaneous generation, 11–16
Stalin, Joseph, 236
Stanier, Roger Y., 75–77, 101–102, 108, 126–130
Staphylococcus aureus, 45–47
Statistical probabilities, 64
Stent, Gunther, 92, 108
Sterilization, spontaneous generation vs., 13
Stewart, William, 45
Streptococcus pneumoniae: Avery's study of, 146–148; capsule and virulence of, 144–146; competence of, 151; Griffith's study of, 142–146; lobar pneumonia from, 142–143; rough (R) vs. smooth (S) strains of, 144–148; transformation in, 142–148, 151; typing of strains, 143–144; untypeable strains of, 144
Streptococcus thermophilus, 174
Streptomyces coelicolor, 83
Streptomycin, 161, 165, 221
Sulcia muelleri, 209–210
Sulfur-reducing bacteria, 57
Superkingdoms, 130
Sutherland, John, 245, 252
Suttle, Curtis A., 32
S value, 76–77
Svedberg, Theodor, 75–76
Swan-necked flasks, 14, 15
Swaziland, fossil record in, 191

TACK superphylum, 206
Tasmanian wolf, 8
Tatum, Edward, 154–160
Taxon(s), 8–9

Taxonomy, 49–52; archaea, 118–119; bacteria, 118–119, 124–132; balance principle in, 111–114; biblical, 7; binomial, 51; definition of, 5; evolution and, 5, 7–8; Linnaeus system of, 50–52; as metascience, 127–128; *The Microbial World*, 101–102, 114; problem with archaea in, 52; problem with bacteria in, 52–60; prokaryote-eukaryote distinction, 129–132; species dilemma in, 51–52; theory-neutral nature of, 49; tripartite division in, 52; virus, 34–35; Woese's analysis and discovery in, 99–115, 118; Woese's primary kingdoms or urkingdoms in, 110–115
T4 bacteriophage, 33, 85–87
Teixobactin, 45–47
Tetracycline, 46, 165, 221
Tetrahymena, 244
Thalassemia, CRISPR and, 176
Thaumarchaeota, 206
Theory, Tree of Life as, 48
Thermophiles, 28, 246
Thermotogae, 218
Thiomargarita namibiensis, 53
Thomson, William (Lord Kelvin), 234
Threonine, 157–158
Thrush, 221
Thymine, 62, 72, 81
Ti plasmid, 188
TNT, microbial degradation of, 55
Tobacco mosaic virus, 33
Tourist diarrhea, 183
Transcription, 91
Transducing particle, 168
Transduction, 167–169, 177; generalized, 168; specialized, 168–169
Transfer RNA (tRNA), 90, 91–92

Transfer RNA (tRNA) synthetases, 179
Transformation, 96, 141–153, 177; artificial, 152–153; Avery's study of, 146–148; competence for, 151; DNA as genetic material in, 142, 147–150; Griffith's discovery of, 141–146, 150; natural, 150–152; nomenclature of, 150
Transforming principle, 147, 154
Translation, 82, 89, 91–92
Tree of Life, 3–9; archaea on, 7, 9, 30, 49, 135–136; arrangement of major groups on, 7; bacteria on, 7, 9, 30, 31, 49, 135–137; Banfield's update of, 136–137; basal radiation of, 135; beginning (origin of life), 232–253; biblical, 4; cellular information and, 16–17; Darwin's concept of, 3, 3–7, 6; dichotomous classification vs., 8–9; domains on, 3, 9; dynamic nature of, 46–47; eukaryotes on, 7, 9, 49; FISH analysis and, 231; fungi on, 7, 9, 21; gene transfer and, 177, 180, 185–186, 188–189, 213–214; Haeckel's, 121–123, *122*; historical path to, 9–16; knowledge gained from, 47; macromolecules and, 72–73; microbes as beginning of, 6–7; microbial dominance of, 9, 21; milestones in developing, 11; molecular morphology and, ix–x; Pace's update of, 135–136; phantom branch of (viruses), 31–35; ribosomes and, 73–74, 78–79; root of, 116–118, 207–208, 214; as scientific theory, 48
Tripartite division, in taxonomy, 52
Triplet code, 86–88; discovery of, 86–87; encoding capacity of, 87–88; redundancy in, 88, 178–179

Tritium, 238
Trypsin, 66–67
Tryptophan, 88, 209
Tyndall, John, 15–16

Ultracentrifuge, 75–77
UPEC. *See* Uropathogenic *E. coli*
Uracil, 62, 72, 243
Urey, Harold, 237–240
Urkingdoms, 110–115, 130
Uropathogenic *E. coli* (UPEC), 182

Vaginal microbiota, 224, 229
Vancomycin, 222
van Niel, C. B., 104, 126–130
Vavilov, Nikolai Ivanovich, 237
Venter, Craig, 134–135
Vertical inheritance, 141
Violence of the Green Revolution, The (Shiva), 43
Virion, 148, 168
Viroids, 33–34
Viruses, 31–35; bacterial (*see* Bacteriophages); bacteria vs., 128–129; discovery and study of, 32–33; DNA, 32, 34; enzymes of, 32, 34; evolution of, 35; "filterable," 33; genetics of, 34; gene transfer via (transduction), 167–169, 177; Lwoff's concept of, 128–129; property of life lacking in, 31–32; RNA, 32, 34, 243–244; shapes of, 33; size of, 33–34; structure of, 32; taxonomy and classification of, 34–35
Vitalism, 13

Waring, Fred, 149
Warm-little-pond hypothesis, 235–240, 252–253; Darwin on, 235; Earth atmosphere and, 236–240;

Warm-little-pond hypothesis *(continued)*
   hydrothermal vents and, 251;
   Miller-Urey experiment on, 237–240;
   progress on plausibility of, 252–253;
   Sutherland's elaboration of, 252
Wasps, *Wolbachia* in, 211
Watanabe, Tsutomu, 164–166
Watson, James, 62, 80–81, 160–161
Watts-Tobin, Richard J., 85–88
Weigle, Jean-Jacques, 170–172
What Mad Pursuit: A Personal View of Scientific Discovery (Crick), 82
Wheat mitochondria, 204–205
Wheelis, Mark, 102, 114–115, 120
White Cliffs of Dover, 25
White flies, endosymbiosis in, 208
White smokers, 246–247
Whittaker, Robert, 123–124
Wilkins, Maurice, 62
Wilson, Allan, 68–71
Wilson, E. B., 201
Winemaking, 22–23, 58, 71
Woese, Carl R., 68, 97, 99–115, 120; *Bacillus* investigation of, 103; belief in data, 111; chloroplast study of, 204; classification of bacteria, 110, 218; death of, 111; discovery of archaea, 106–115, 130; Mayr's praise and doubts of, 112–113; methanogen study of, 102–115; methodological impact of, 132–137; objection to prokaryote terminology, 130, 131; proposed mechanism of ribosomal movement, 101, 108; publicity on "third form of life," 109–110; on RNA's capacities / potential, 244–245; sensitivity to oversight, 108–109; sequencing method of, 99–100, 118, 132–133, 219; taxonomic focus of, 99, 118
"Woese's Angels," 130
"Woese's Army," 130
Wolbachia, 211–212
Wolfe, Ralph S., 102–109, 118; culturing of methanogens, 104–106; reaction to publicity on "third form of life," 109
Wollin, Eileen, 105–106
Wollin, Meyer, 105–106
Wood, Tom, 80
Wright, Sewall, 237

X-ray diffraction, 61–62
Xylella fastidiosa, 209–210

Yeasts, 21–23; *Candida albicans* in human microbiome, 222; cultivation of, 21; fermentation by, 22–23; metabolism of, 198–199; *Saccharomyces cerevisiae*, 22, 77, 198–199; as study organism, 21–22

Zimmer, Carl, 136
Zinder, Norton, 167–168
Zoology, 9
Zuckerberg, Mark, 176